# Diary of an
# Eco-Builder

## Will Anderson

Green Books

First published in 2006
by Green Books Ltd
Foxhole, Dartington
Totnes, Devon TQ9 6EB
www.greenbooks.co.uk

All photos © Will Anderson 2006 except for
pages 1, 7 & 24 © Andy Paradise
page 195 © Adam Lee

All architectural drawings © Peter Smithdale, Constructive Individuals

Design by Julie Martin

Printed by Butler & Tanner, Frome, Somerset, UK
on 100% recycled paper

British Library Cataloguing in Publication Data
available on request

ISBN 1 903998 79 4

# Contents

# Acknowledgements

All the entries in this diary were first published in the Property section of *The Independent*. I owe the paper and my editors, Bernice Davison and Madeleine Lim, a debt of gratitude for giving me the opportunity to share the experience and let my ideas off the leash. I would also like to thank Amanda Cuthbert and John Elford of Green Books for their unfailing encouragement.

Thanks to all the professionals, tradespeople, craftspeople and labourers who contributed in so many ways to the creation of Tree House, above all to our architect Peter Smithdale, principal builder Steve Archbutt, contractor Martin Hughes and engineer Ian Drummond. Thanks also to the many suppliers who supported the project and to the Norwich and Peterborough Building Society whose enlightened attitude to ecological buildings is in marked contrast to the mass of lenders whose interests extend no further than 'standard construction'.

Thanks to all our friends, family and neighbours for their support and help, especially to Sara Maitland who gave me the confidence to write the story. Finally, thank you to Ford, who keeps me on the rails and makes everything possible.

*Will contemplates the dream.*

# Introduction

Every year in late September architecture enthusiasts of all persuasions get the chance to over-indulge during the London Open House weekend. For two days only, buildings of all kinds are opened to public scrutiny across the capital, including many private houses.

In September 2002, Ford and I took the opportunity to explore the festival's offering of ecological homes. Although the buildings we saw were very diverse, each was inspiring and interesting in its own way. Whatever their individual pros and cons may have been, each had the merit of existing, of having been built. This simple fact was enough for me to temporarily put aside my assumption that our own self-build dream was only possible if we left London. The very next day, I began my search.

I did the obvious things first. I bought the self-build magazines and checked out their listings of building plots. I signed up to the online databases. I contacted the council's auctioneer. In a short space of time I covered a lot of ground and got precisely nowhere.

The next day I did what any self-respecting property-hunter would do first: I checked out the local estate agents. As I wandered down Clapham High Street my simple enquiry – "Do you by any chance have any land for sale?" – was met with surprise, incredulity and barely disguised derision. There were two agents in Clapham Old Town but I turned for home after the predictable disappointment at the first, safe in the knowledge that my afternoon had been wasted. But the urge for completion overtook me and, in a moment that in retrospect holds a perfect dramatic intensity, I turned back.

"Land, sir? As a matter of fact a building plot has just come in. A bit tricky though – there's a bloody great tree in the way."

* * * * *

This diary begins two years later, just before work started on Tree House. Our story is a very personal one, rooted in a specific site, vision and set of personal priorities. Nonetheless, the issues we tackled are universal and I hope that our story has a resonance for anyone with an interest in building, design and ecological living.

Our inspiration was and is our tree. Rather than look to standard practice and seek to do better, we set our sights on the best possible example of environmental design and sought to meet it. Tree House is the result.

*The plot, only six metres wide, is completely dominated by the tree.*

## 01 SEP 04

**My diary began just before the builders arrived. It was a very exciting time: almost two years had passed since I had first set eyes on the building plot. Now, at last, we were ready to turn our carefully prepared eco-dreams into reality. I relished the September sunshine, cherishing the thought of waking to the same rays in the bright rooms of Tree House twelve months later. Little did I know that the following September we would still be living in the dim light of our cramped Brixton flat, the build of a lifetime far from complete. The vision, however, never faded.**

# A house that works like a tree

I'm standing in a small patch of neglected land in a quiet corner of Clapham, South London. A tiny urban wilderness of brambles and bracken. Ignored by humanity, it is home to robins, foxes, squirrels and innumerable scurrying invertebrates. The horizontal extent of the plot is less than 150 square metres but it is dominated by the vertical presence of a mature multi-stemmed sycamore, a towering structure that reaches up and out to the sky. Here, on a bright August day, the overgrown plot is transformed by the ever-changing light in the tree canopy and the gentle sway of the branches in the breeze.

The biodiversity of the smallest patch of land is dramatically improved by a tree. Recent research by the University of Sheffield into the density of bugs in urban gardens found that planting a tree is one of the best ways of encouraging local wildlife, especially invertebrates. You are effectively laying the foundations for a grand condominium of urban flora and fauna, a luxury habitat that protects its tenants from the ravages of the elements and provides them with a rich supply of nutrients. Everyone benefits, from the beetles that burrow in its bark to the millipedes that forage in the decaying organic matter at its base.

Rather different foundations are planned for this plot, however: deep shafts of concrete, driven into the untouched earth. For this little patch of ground, never built on in modern times, is now a building plot with full planning permission for a three-storey house. Any day now the piling rig will arrive and the local ecosystem will be ripped up. My partner and I have set this process in motion: the house that will rise here will be another small step for man, reflected a million-fold across the

The architect's model
of Tree House.

country as the government embarks on its giant leap to overcome the housing crisis.

The government badges its building programme as creating 'sustainable communities'. Yet it is fairly obvious that building is still far from sustainable. Not only does house-building destroy local wildlife habitats, it also consumes natural resources and creates little engines of ongoing consumption, above all of carbon-intensive energy. Buildings account for half of our energy use, houses for about a quarter, so they are critical to the task of tackling climate change and making our world sustainable. Yet current building practice is a long way from 'meeting the needs of the present without compromising the ability of future generations to meet their own needs' – the original definition of sustainable development in the 1987 Bruntland Report, *Our Common Future*. Every new house is a new burden on the environment that future generations may pay for dearly.

There is another way. The house we are planning for our little slice of Clapham will not place any burdens on the future. Despite those concrete foundations, it will be built primarily with low-impact, non-toxic materials. The principal construction material, timber, will be sourced from well managed forests. Water consumption will be cut to half the typical rate. Waste will be driven near to zero. The immediate natural environment will be protected and enhanced. Above all – and this is the really tricky bit – the house will be so energy-efficient that all our power and heating needs will be met through our own renewable energy generation, on-site. We will be self-sufficient in energy and so free from fossil fuels. Arguably, this will be the most 'environmentally-friendly' contemporary urban home in Britain.

But that's only half the story. Our number one priority is to build a house that is a wonderful place to live. We want to make contemporary environmental design work for us as well as for the planet. If 'green living' is all about sacrifice, forget it. We want more: more light, more comfort, more beauty, more health and more style. I am confident that a holistic approach to environmental specification will deliver a quality of life for us that is far superior to that offered by a gas-guzzling design.

Back at the plot, my ecological ambitions seem hard to square with the imminent arrival of JCBs and concrete trucks. But with care, we can enhance the biodiversity of the land over the course of the build. The house will only occupy a third of the plot, so we have space to improve upon the brambles and bracken. There will be a small but richly planted organic garden full of flowers that attract bees, butterflies and other insects. There will be nesting sites for birds, bats and solitary bees and wasps; an invertebrate 'creature tower' made from left-over building materials; and a slowly decaying log-pile. A pond will provide an entirely new environment and hopefully attract a range of new species. Above all, the house has been designed to protect the tree so that the human contribution to the plot's biodiversity can thrive alongside the rest of the local wildlife for years to come.

The house will in fact work like a tree. It will be a striking timber structure – beautiful, strong and adaptable – that harvests all its energy from the sun and provides a secure and valued habitat. I may be an incurable romantic, but looking up into the branches today, I am hopeful that whatever difficulties beset me during the build, life in Tree House will be worth every bit of our ecological ambition.

*The front elevation of the house is full of movement, evoking the organic liveliness of the tree itself.*

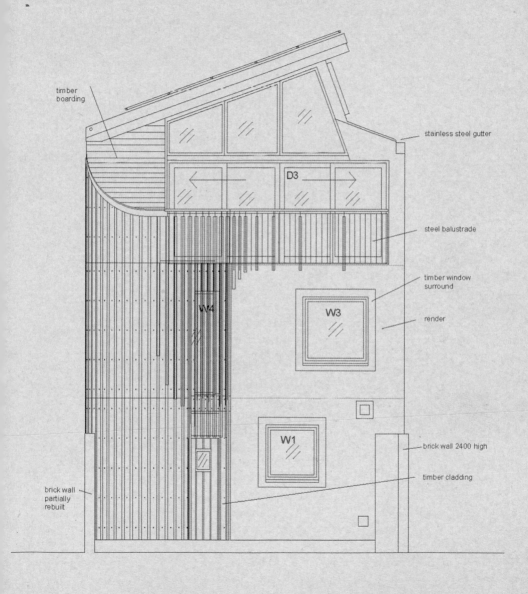

timber
boarding

stainless steel gutter

D3

steel balustrade

timber window
surround

W4

W3

render

W1

brick wall 2400 high

timber cladding

brick wall
partially
rebuilt

13

## resources

There are many other ecohouses in Britain ranging from earth-sheltered rural dwellings to high-density urban developments. For inspiration, see:

The Beddington Zero (fossil) Energy Development (BedZED). A ground-breaking development in south London by BioRegional, an organisation with a truly holistic approach to social environmental design (www.bioregional.co.uk).

The Brampton Ecohouse. A demonstration project to show that low-impact houses can be affordable, attractive and easy to build and maintain (www.brampton-ecohouse.org.uk).

The Brithdir Mawr Community. A collective in Wales seeking a sustainable relationship with the land and local environment (www.brithdirmawr.co.uk).

Findhorn. A long-established ecovillage with many different building types (www.ecovillagefindhorn.org).

The Hockerton Housing Project. A remarkable low-impact development in a Nottinghamshire field (www.hockerton.demon.co.uk).

The Leicester Ecohouse. A show house full of information and advice about environmental issues in the home (www.environ.org.uk/ecohouse).

The Nottingham Ecohome. A thorough renovation of a semi-detached Victorian villa (www.msarch.co.uk/ecohome).

The Underground House. An earth-sheltered house built in a disused Cumbrian quarry (www.theundergroundhouse.org.uk).

The Yellow House. A renovation of a terraced house that leaves no ecological stone unturned (www.theyellowhouse.org.uk).

### Publications

*Ecohouse 2* (S Roaf, M Fuentes and S Thomas, Architectural Press 2003). An extensive introduction to ecobuilding, with a focus on the Oxford Ecohouse.

*The New Autonomous House* (B & R Vale, Thames and Hudson 2000). A detailed technical guide to an influential house in Nottinghamshire.

*Sustainable Housing Schemes in the UK. A guide with details of access* (N White, Hockerton Housing Project 2002). Information on a wide variety of individual houses, community projects and developments.

*The Woodland House* (B Law, Permanent Publications 2005). A blow-by-blow account of the construction a remarkable house in the woods.

*Dwell Well* promotes the development of affordable green homes and settlements in Britain (www.dwellwell.org).

# *29 SEP 04*

The building of Tree House began on September 20th 2004. After so many months of preparation and anticipation we simply had to get going.

The first man on site was Steve Archbutt, site foreman. We greeted each other with equal amounts of politeness and reservation. It did not take me long, however, to realise how much of an asset Steve would be to the whole project: thoughtful, straightforward, capable and meticulous. Steve would keep me sane through the long months ahead.

| *The digger arrives and the clearance begins.* | *The site is rapidly laid waste.* | *The setting out begins – a few lines in the earth, promising much.* |

# Climate-sensitive design

Today the first sod was turned. It is precisely two years since I first set foot on the tiny patch of urban wilderness that is now our building plot. I hope the coincidence is auspicious and that the delays to our project have at least brought the planets into supportive alignment.

The latest hitch arose when our contractor's insurers quoted a premium four times higher than his estimate. Lloyds' underwriters may be nervous about groundworks in London's shifting sands but our expensive piled foundations ought to make the house more robust than trenches of concrete. If the insurers can't be persuaded, our budget will also be sinking deep into the brown.

But this is not a day for financial nail-biting. Up to now, the pleasures of self-build have been entirely cerebral. Finally I am enjoying the more elemental experience of purging and renewal as our overgrown, rubble-strewn plot is stripped back to the naked earth, ready for the building to begin. All that remains is the tree that dominates the plot and a great empty volume of air where our house will grow.

When I first confronted this big three-dimensional nothingness, the complexity of a house – any house – suddenly seemed overwhelming. Thousands of choices had to be made to mould the emptiness into spaces, structure and details, yet each choice seemed to beg a thousand questions.

I quickly discovered that design is not a beautifully rational process but a muddling through in which competing interests rise and fall with the phases of

*design detail*

## Roof windows

The solar energy reaching the southern roof of Tree House is transformed into hot water and electricity. The indirect solar energy that arrives at the northern elevation also makes an energy contribution by bringing daylight into the study at the top of the house.

The study has a line of five VELUX roof windows positioned between

the spreading branches of Douglas fir that hold up the main roof. The only northern windows in the house, they provide an ideal light to work by. They can also be opened in the summer to turn the whole building into a natural ventilation shaft driven by rising warm air.

tho moon. In this mix, the constraints of the site were invaluable in forcing practical form upon the ideas and dreams that we brought to the project. We took very seriously our responsibilities to protect the tree, to meet the conditions of the planners and to respect the interests of our neighbours. Another site characteristic, often ignored by contemporary architects, was particularly important to us: the climate.

Climate sensitivity was once the starting point for all building. Across the world, vernacular houses are designed to exploit the warmth and light of the sun, the cool earth and the shelter of the land. In the era of cheap fossil fuels, all this was forgotten as architects could create comfortable interiors regardless of the external environment. Victorian builders, like developers today, replicated their pattern-book villas willy-nilly, knowing that a plentiful supply of cheap fuel would keep their buyers happy.

If we are to reduce our dependence on fossil fuels, we need to rediscover climate-sensitive building. The Beddington Zero (fossil) Energy Development (BedZED) in Sutton, south London, is an impressive attempt to do this. The terraced houses are a contemporary take on 'passive solar' design, using huge south-facing windows and thick concrete walls to trap and store the sun's energy. Architect Bill Dunster describes this as the 'warm cave' approach, acknowledging the climatic wisdom of our earliest homemaking forebears.

### *profile*   Steve Archbutt

Steve claimed that his ancestor, Samuel Archbutt, had been the principal builder to Thomas Cubitt, the developer of Clapham Park. As our slice of land originally formed part of the back garden of one of Cubitt's mansions, it seemed that Steve's appointment as site foreman for Tree House was a highly appropriate continuation of a family tradition.

Steve came to the project with lots of experience of unusual and challenging builds, including his own timber-frame home. Tree House had some surprises in store even for Steve but he tackled them all with great care and an overriding commitment to quality. He also succeeded in sustaining the good humour and good will of everyone on site throughout the build, an invaluable achievement that was all too easy to take for granted.

As our plot has no southern aspect, we cannot build a warm cave, although big windows to the east and west will bring energy-saving daylight deep into the house. We can however pitch our roof to the south and harvest solar energy with technology: solar panels and photovoltaic (PV) modules. Government grants will ease the installation of both.

Solar panels are radiators in reverse: the sun heats the panel which then heats the water piped across it. PV modules are at the other technological extreme, using sophisticated semiconductors to turn light into electricity. As their efficiencies improve and costs fall, PVs offer real hope for large-scale but unobtrusive renewable power generation.

If you are building a house or an extension, include a roof with a southern pitch. If you have a south-facing roof already, consider its potential. You don't need blazing sunshine for solar hot water: even in gloomy years you can get at least half of your hot water from your roof. With PVs, you can generate a decent slice of your power needs as well, selling to the Grid when you don't need it and buying it back when you do. We aim to go all the way: by specifying an ultra-efficient house, we will meet all our needs from our little roof-top power station.

A climate-sensitive house will let solar heat in when it is needed and keep it out when it is not. Shading from the environment, such as deciduous trees, or from building details, such as deep eaves, is critical. Avoid building a home on the top of a hill where the wind will whip away the heat and keep your fuel bills high. Look for shelter from the landscape, trees and existing buildings. For the greatest protection, build into the side of a hill.

Closing the gate on our newly turned plot, my thoughts turn to the history of the site. The plot was once part of the garden of a grand villa in Thomas Cubitt's nineteenth-century Clapham Park development. It doesn't take me long, following

*The architect's 3D computer model reveals the priority of solar energy in the design of the house.*

the block round, to reach one of the remaining houses from his ambitious scheme. It looks well-built – brownie points for sustainable resource use – but its structure is defined by the chimneys that rise within to the crowded ceramic pots on the roof. This house was built for people who could afford plentiful coal and plentiful servants to shovel it.

The windows at the rear of Tree House will repeat the pattern of the fenestration of Cubitt's mansions in a playful acknowledgement of the architectural history of the area. But there will be no chimneys on Tree House, no fireplaces and certainly no servants. The time is surely ripe to rethink our own servitude to architectural styles that are no longer fit for the environmental demands of the new century.

## resources

For information on grants for solar panels and photovoltaic modules, contact the Energy Saving Trust (0800 915 7722, www.est.org.uk).

Beddington Zero (fossil) Energy Development: www.bioregional.co.uk.

### Design detail
Roof windows supplied by the VELUX company (0870 405 7700, www.velux.co.uk).

### Publications
*Building for a Future*. A highly informative quarterly magazine published by the Green Building Press (www.buildingforafuture.co.uk). The GreenPro database of ecobuilding advice and information is also accessible from the website.

*The Climatic Dwelling* (EO Cofaigh, JA Olley and J Owen Lewis, James and James 1996). Demonstrates how vernacular buildings from across the world cope with the local challenges of climate and terrain.

*Planning for Passive Solar Design* (Energy Efficiency Best Practice Programme ADH010). Free to download from the Carbon Trust (www.thecarbontrust.co.uk).

*Shelter* (L Kahn and B Easton, Shelter Publications 2000). More global inspiration: hundreds of vernacular buildings.

*Solar Energy and Housing Design* (S Yannas, Architectural Association 1994). A detailed guide to the many ways houses can benefit from the light and heat of the sun.

# 27 OCT 04

Before we could build anything ourselves we had to prop up our neighbours – the first house in a terrace climbing a shallow but significant hill. This was technically tricky, requiring the insertion of temporary sheet piling along the boundary. Once this was done we began the installation our first piece of serious eco-technology: the ground loop for our heat pump.

Most ground-source heat pumps draw their energy from pipes laid in trenches in long gardens. As this was not an option in our dinky courtyard garden, we drilled some boreholes instead. The pipes were put down the boreholes over a year before the heat pump was switched on for the first time. They are now hidden forever beneath the house.

## Heat pumps

Why do cats sit on fridges? Fitted kitchens have banished this nesting site from the domestic routine of most cats but I remember, as a child, the imperious look of Moppet, sitting high up on the fridge-freezer in a chilly Edinburgh kitchen. It seemed a strange choice to me, given her fondness for lying dangerously close to the gas fire.

Cats are, of course, smarter than little boys. They understand that in order to keep the inside of a fridge cold, heat regularly has to be pumped out of it. This heat is concentrated and released at the back of the fridge, creating a warm draught for furry hides above.

There are four cats in our current family, all experts in finding the warmest corner of our south London flat. Little do they know, but two

*The temporary sheet piling is installed to protect the neighbours from our digging.*

blocks away at our Clapham building plot, a piece of kit is currently being installed that will make our entire house feel like the top of a fridge-freezer.

The ground has been well prepared: the last two weeks on-site have been dominated by earthmovers, carving a tentative house footprint in the clay. Our four young (male) neighbours have been awestruck. Little Lucas is so gripped by the events on the other side of his garden fence that he bursts into tears if 'Mr Digger' stops for a fag. Happily, Mr Digger – otherwise known as Steve – has not dug up anything other than soil, gravel and clay. There are no signs of Roman remains, unexploded bombs, Victorian wine cellars or anything else that might obstruct our piled foundations.

We are using piles, rather than strip foundations, in order to protect both the roots of our beloved tree and the house itself, should the earth move beneath it. But we're getting some added value from the piling rig: four 25-metre boreholes. The pipe that we're threading through each borehole will transport a refrigerant that will take the heat out of the ground and into our house in exactly the same way that heat is pumped out of a fridge.

A 'ground-source heat pump' is highly unusual in central London, although the invertebrates in London Zoo have enjoyed one for some time. This is not because the technology is unproven – they are installed in most new homes in Sweden – but because gas is cheap. The renewable energy under your back garden is there for the taking but you still need electricity to pump it out. Although a heat pump will generate about three times as much heat energy as the electrical energy you put in,

*Mark the piler steers his rig into place.*

*George and Steve thread the refrigerant pipes down the boreholes.*

electricity costs three times the price of mains gas. Not surprisingly, the UK heat pump market is concentrated where mains gas is not available.

We are installing a heat pump because we care about environmental costs as well as our monthly bills. The carbon emissions generated by a heat pump vary depending on how the electricity is generated but over the year a ground-source heat pump will be greener than even the most efficient gas boiler. In order to achieve our 'zero carbon' goal, our heat pump will be driven by our roof-top solar power station.

Sixty-two percent of the energy consumed by households goes on space heating, so it's a good place to begin any domestic eco-improvements. Your first priority should always be to stop the heat escaping: an afternoon spent draught-proofing will, I promise, transform your life. Improving insulation is rarely a top DIY priority but, even if you can't see the result, you will feel it.

If you are ready for a new heating system, get advice about the most energy-efficient and climate-friendly options for your circumstances. If you're after a new gas boiler, make sure it's a condensing boiler that extracts energy from its own exhaust gases. If you're reliant on oil, coal or electricity, it's certainly worth considering a heat pump. If the capital cost seems a little steep, remember that heat pumps have low maintenance costs and a very long working life.

Wood, the fuel of choice for almost the entire history of humanity, remains a highly ecological fuel as growing and burning wood does not result in any net increase in

The boreholes are backfilled with bentonite, a highly conductive clay.

Finally, the pipes are pressure-tested. Thankfully, they got the all clear.

atmospheric carbon dioxide. Automated wood-pellet boilers are the most advanced carbohydrate burners and, like ground-source heat pumps, attract government grants.

In an exceptionally energy-efficient home, 'casual' heat sources can also be important. In *The New Autonomous House*, ecological architects Brenda and Robert Vale contemplate populating their bedroom with fifteen cats to meet their small heat demand. A cat produces about 15W of heat, so we can rely on a steady output of about 65W (including an extra 5W for alpha-male Trevor). This is significant, though we know from our nightly experience of being pinned to the bed that the cats value our heat more than we will ever value theirs.

Hopefully when Tree House is complete and occupied, Trevor and family will no longer feel the need to cling to their guardians but instead stretch their considerable bellies over the warm floors, enjoying Moppet's top-of-fridge experience in every corner of the house.

## Chimney pots

Tree House boasts four towering clay chimney pots, but they belch no smoke. As the ultra-efficient house is heated by solar energy stored in the ground, we have no need for a fireplace. Consequently, our garden is our hearth and the salvaged chimneys provide a tongue-in-cheek classical frame for the fiery planting at the centre of the garden.

## profile

### George Archbutt

George, Steve's nephew, contributed to just about every job in the building of Tree House from cutting back the brambles to putting the roof on. He was probably happiest nailing timber at the top of the house but he didn't complain when he had to descend into the sewer to connect our drain. George was also the only man on site who could be trusted with a theodolite (he knew a straight line when he saw one).

## resources

The ground-source heat pump installed in Tree House was supplied by Ice Energy Heat Pumps (01865 882202, www.iceenergy.co.uk).

The integration of the heat pump with a roof-top solar thermal panel was designed by Ice Energy Scotland (0845 600 1020 www.iceenergyscotland.co.uk).

For information on the grants available for domestic heat pumps and wood pellet boilers, contact the Energy Saving Trust (0800 915 7722, www.est.org.uk).

The Ground Source Heat Pump Club is a trade organisation providing news and information about the technology (www.gshp.org.uk).

### Publications

The following Best Practice in Housing publications are free to download from the Energy Saving Trust website and can be ordered via the helpline:

*Domestic Ground Source Heat Pumps: Design and installation of closed-loop systems* (CE82/GPG339).

*Heat Pumps in the UK – a monitoring report* (GIR72).

Other Best Practice in Housing publications describe the most effective ways of heating your home with gas (CE30), oil (CE29), electricity (GPG345) and solid fuel (CE47).

# 24 NOV 04

Our ground works took forever. Concrete piles were drilled and filled amid the boreholes, then the slab was prepared and poured on top before the final stitching-in of the retaining wall and pond. We spent so much time in a field of mud and concrete that my 'ecobuilder' identity began to feel a little shaky. However the amount of concrete used overall was not excessive as the main structure of the house was made out of timber.

Given our setting – next to a tree, on a hill, below a terrace – we had few qualms about laying concrete foundations. When our Brixton flat began to split apart under the influence of a tree in the back garden, we felt sure we had made the right decision.

*Steve contemplates the cement garden.*

# Construction materials

A new garden is flowering on our ravaged Clapham building plot. Thick grey stems have sprouted in the bare clay, from which rough tendrils rise to the sullen November sky. A vigorous grey mould spreads between them, brittle underfoot. Our great sycamore appears to be shedding its leaves in pure disgust.

There's nothing like concrete to inspire fantasies of ecological horror, not least because the manufacture of cement, the key binding ingredient, is a major contributor to climate change. Cement is made by roasting chalk or limestone in very high temperature kilns, a chemical process that produces carbon dioxide and uses lots of carbon-intensive energy. Around one kilogram of carbon dioxide is emitted for every kilogram of cement produced.

It doesn't help that concrete is also strongly associated in the British imagination with dismal post-war housing estates, multi-storey car parks and the Hayward Gallery. And it's just such unpleasant stuff: the truck that has been filling our piles has no redeeming features, it is a diabolical beast that lifts its stinking tail and evacuates a long stream of toxic diarrhoea into the wheelbarrows waiting below.

But is concrete really so evil? After all, there are lots of eco-builders who think it's the best thing since wholemeal bread. Whatever the downside to the manufacture of cement, concrete's contribution to the performance of a building means it can still be part of a green specification. The case for concrete rests on its strength and on its heat-retaining properties.

We are using concrete for strength: although most of Tree House will be built from timber, our roots must be strong enough to sustain the house over a long lifetime, potentially beyond the life of the mature sycamore immediately next to it. Longevity should be key to any ecological specification: if a building performs well over a very long life, the ecological costs of construction materials pale into relative insignificance.

*Steve and George in front of the sleeves for the concrete piles.*

*After its insertion in the hole, Mark secures the sleeve before the concrete pour.*

We recently visited the site where concrete may have been poured for the very first time in these isles: Richborough, the Roman gateway to Britain on the low Kent coast. Today, after almost 2,000 years, the remarkable walls of Richborough fort still stand tall. Although the stones of the once great ceremonial arch have long been plundered, the concrete foundations remain. The Romans were adept at exploiting the strength and versatility of concrete, although the survival of their most famous concrete building, the great light-pierced dome of the Pantheon in Rome, may say more about its beauty than its structural integrity. In general, ugly buildings do not survive; truly sustainable buildings must be beautiful as well as strong.

*The concrete truck delivers its load.*

Concrete walls and floors are also used in green buildings to smooth out temperature swings by absorbing and slowly releasing heat. In hot summers, this 'thermal mass' helps to prevent people rushing home to put their feet up in front of the power-hungry air-conditioner. When used in combination with big south-facing windows, exposed concrete can also keep a house warm for longer into the evening, potentially reducing heating costs.

There are, however, other ways of preventing overheating, such as good shading, so it is possible to build robust lightweight buildings out of more sustainable materials. Buildings made from timber have the lowest environmental impacts in manufacture because timber is a renewable resource. The sustainable harvesting of timber also extracts carbon from the atmosphere and locks it up in the structure of the building.

More eco-friendly heavyweight materials include stone, reclaimed bricks, cob or even old tyres filled with earth and rubbish – the signature construction material of 'earthships'. Straw bales are a versatile lightweight material for cheap, well-insulated walls.

In another 2,000 years time, I don't suppose Tree House will be standing, despite its concrete foundations. Hopefully, the house will have been carefully dismantled and recycled, each component becoming a nutrient for a new building or product. Perhaps the timbers will be re-used in another sustainable building in an era when such design is standard practice. If things don't work out so well, they might yet be used for flood defences or latter-day arks. Perhaps, in 4004, visitors from the balmy new capital of Snowdon will visit the low, mosquito-infested swamp of London and marvel at the longevity of the ancient city's rust-flamed concrete roots.

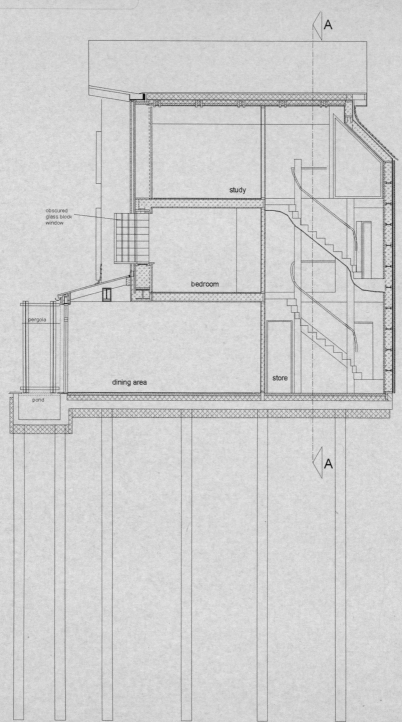

obscured
glass block
window

study

bedroom

pergola

dining area

store

pond

A

A

## Garden walls

Our building plot was fronted by high brick walls, overrun with creepers and looking a little worse for wear. As they were structurally very weak, they had to be pulled down and rebuilt. Fortunately they had predominantly been built with lime mortar so it was possible to reclaim many of the bricks, avoiding the significant energy costs embodied in new bricks.

The new walls were made from a combination of our own bricks and similar stocks from a pub renovation in New Cross. In time the dappled pattern of the rich brickwork will once again be overrun with ivy, a core ingredient of the wildlife garden at the front of Tree House.

## The piling team

Our concrete piles and 25-metre boreholes gave the piling team plenty to do. Although their task involved plenty of hard labour shifting London clay, it also required precision and care to ensure that the holes did not collapse before the job was done.

Despite spending every day trudging around in the mud, digging holes and pouring concrete, the piling team remained remarkably cheerful throughout the job.

*Compared to the boreholes, the piles were quite shallow. Compared to the house, they seemed extravagantly deep. This house is not going anywhere.*

## resources

To explore the potential of sustainable construction materials, contact eco-builders'-merchants such as Construction Resources (020 7450 2211, www.constructionresources.com) or the Old House Store (0118 969 7711, www.oldhousestore.co.uk).

For details of Earthships in Britain see www.sci-scotland.org.uk and www.lowcarbon.co.uk.

### Publications

*Beddington Zero (Fossil) Energy Development Construction Materials Report* (BioRegional Development Group 2003). Detailed analysis of the materials specification for a radical eco-development that used large amounts of concrete.

*Building with Cob* (A Weismann and K Bryce, Green Books 2006). A new guide to a very old approach to building with earth and straw.

*Building with Straw Bales* (B Jones, Green Books 2002). Everything you need to know about this cost-effective and energy-efficient building material.

*Ecohouse 2* (S Roaf, M Fuentes and S Thomas, Architectural Press 2003). Includes a substantial section on the environmental impact of building materials.

*The Green Building Handbook* (T Wooley, S Kimmins, P Harrison and R Harrison, E & FN Spon 1997). Comprehensive information on the environmental impacts of building materials.

*The Green Guide to Specification* (J Anderson and D Shiers, Blackwell Science 2002). A highly systematic guide designed to give professional specifiers quick answers.

*Reducing Overheating – a designer's guide* (Energy Efficiency Best Practice in Housing CE129). Includes a section on thermal mass. Free to download or order from the Energy Saving Trust (0800 915 7722, www.est.org.uk).

*Timber Building in Britain* (RW Brunskill, Cassell 2004). A detailed technical and historical overview of the various forms of timber building.

*The Whole House Book* (C Harris and P Borer, Centre for Alternative Technology 2005). A thorough guide to eco-building that gives special attention to material.

# 15 DEC 04

As the winter set in and the days darkened, Steve and George kept their heads above the mud of the building site with consistently good humour. We were lucky that our ground works coincided with an unusually dry winter but the mud still got everywhere.

The slab was finally poured into its shuttering over the concrete piles on Christmas Eve. We missed this long-awaited moment, so Ford and I returned from our Christmas holiday to find the mass of dirty-brown clay transformed into a gleaming grey platform from which our dreams would rise. Our rainwater cistern was installed earlier in the month: another labour of concrete love.

*Our Christmas present from the site team – the slab.*

# Rainwater harvesting and water efficiency

What's the definition of a self-builder? Someone who puts brick on brick in all weathers, huddling in a caravan at night in the sure knowledge that the house will never be ready for Christmas? Or someone who dreams up a fantasy home but leaves the hard work to a contractor, only visiting the site to get in the way and give the foreman something to complain about?

I confess that I am closer to the second description than the first, though our site foreman, Steve Archbutt, is unfailingly cheerful. This week, however, I finally got my hands dirty. Paying my morning visit to the site, I found Steve and four other men sinking into the mud as they struggled to shift the concrete sleeves for our underground rainwater tank. I couldn't just stand there and watch their pain, so I was rapidly – and literally – roped in.

Life on a building site in London is a truly multi-cultural experience. Did you hear the one about the Englishman, Scotsman, Russian and Albanian trying to build a concrete water tank without a crane? The Englishman gave the orders, the Albanian complained bitterly, the Russian ignored them both, and the Scotsman (that's me) did his liberal best to respect all their opinions in order not to lose any fingers. Five hours and a lot of pan-European swearing later, the tank stood solidly in its hole and Steve assured me that nothing in the entire build would be as tough.

Rainwater collection, or 'harvesting' as it is romantically called, makes good ecological sense for two reasons. First, it eases the pressure on the mains water supply, reducing upstream energy and environmental costs. Last year English

*Steve digs the hole for the rainwater cistern.*

*George checks the depth of the cistern chamber.*

Nature reported that 160 wetland nature reserves were in danger of drying up because of ground-water extraction. Secondly, it reduces the risk of flooding during storms by buffering the deluge before it hits the drains. This is an increasingly important function: images of English villages under water are becoming familiar, and earlier this year thousands of fish died when London's sewers were overwhelmed, flushing raw sewage into the Thames.

Rainwater is a great resource, but the more of it you want to use, the more kit you need. To provide for your garden, all you need is a tank or water butt. To use water inside your home, you will need a much bigger tank, a completely separate set of pipes from those carrying mains water, plus an electric pump and some filters. To do anything beyond flushing toilets, and possibly clothes washing, you will need chemical or biological filtration as well.

So what's the most ecological choice? The answer is not obvious, as the more complex the kit, the greater the ecological costs of the system – pump energy, maintenance and capital depreciation all have hidden burdens. If you are ultra-efficient in your use of water, the savings of full-scale rainwater-harvesting may be meagre. The sums look much better for groups of houses, large developments and commercial settings (where most of the water goes down the toilet and bubble baths are rare). The more complex approaches also require maintenance and therefore commitment. If you don't clear your gutter regularly, your toilet will begin to smell of rotting vegetation. This is a significant issue, for if ecological design is to become universal, it must be robust: it must work for everyone, not just the committed.

Our rainwater tank will replenish our pond, water our garden and keep us going in emergencies. Rather than pumping rainwater into our home, we are using good design inside to minimise our water consumption. Our two toilets will be state of the art and our appliances will be ultra-efficient in both energy and water use. Our taps will be fitted with aerators, which give you a fuller flow for less water, and flow regulators, which prevent high-pressure systems losing gallons of water whenever you turn a tap on. The brand we have specified, Hansgrohe, incorporates both of these technologies in its bathroom taps as standard. In Germany, everyone's water is metered and prices are much higher, so it's not surprising that it's a German company leading the way here (happily German design looks great too).

In the festive spirit, here's a quick damp quiz to round off my building year.

1. What's the most water-stressed city in Europe? A) London, B) Madrid, C) Istanbul. 2. How much of our drinking-quality tap-water is actually used for drinking or cooking? A) Less than 1%, B) 5%, C) 20%. 3. Will we be showering in Tree House by Christmas 2005? A) Without the slightest doubt, B) You can tell he's a rookie, C) Dream on. The answer, of course, is A in every case.

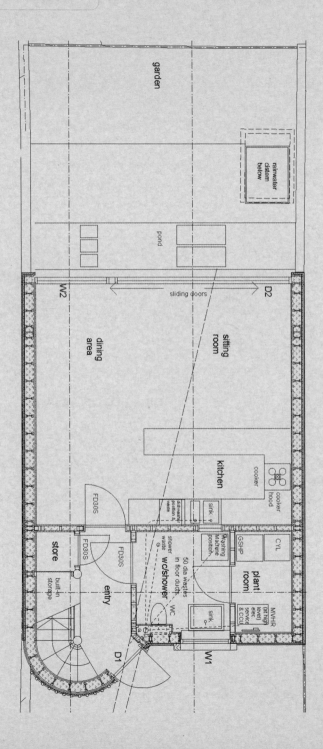

## design detail

### Ifö Cera ES4 toilet

The Ifö Cera ES4 toilet is the epitome of robust ecological design. It is the most water-efficient toilet on the market yet it doesn't even have a dual flush. Dual flush toilets make sense but they have their problems: choose the wrong button and you may have to flush again. Furthermore, because the dual flushing mechanism uses valves rather than the traditional siphon, the toilet is more likely to jam or leak in the long term. A toilet that quietly leaks into its bowl is easily ignored but can waste gallons of water every day.

The ES4 is a single flush siphonic toilet that can never leak and it uses only 4.5 litres per flush. It can neither be misused nor go wrong without you noticing. The make is Swedish but it was designed by the British company, Elemental Solutions.

## resources

The taps and showers for Tree House were supplied by Hansgrohe (0870 7701972, www.hansgrohe.co.uk).

For details of suppliers of rainwater harvesting systems, contact the UK Rainwater Harvesting Association (www.ukrha.org).

### Design detail
Ifö Cera ES4 toilets supplied by the Green Building Store (01484 854898, www.greenbuildingstore.co.uk).

### Publications
*Conserving Water in Buildings*. Eleven concise fact sheets on water efficiency in homes, available from the Environment Agency (08708 506 506, www.environment-agency.gov.uk).

*The Water Book* (J Thornton, Centre for Alternative Technology 2005). A useful guide to the sustainable ways of sourcing, treating and using water.

*This plan of the ground floor and back garden shows the rainwater cistern beyond the pond that divides the house from the garden.*

# 26 JAN 05

With the slab in place, I was confident that we would soon see the house rise from its exceptionally solid foundations. There were, however, a few nips and tucks to the ground works still to complete. The biggest tuck of all was the pond: I couldn't believe how much work went into building it.

With hindsight, it seems a particularly extravagant feature of the design. If we knew at the beginning what our costs would be at the end, it would definitely have been cut. Thankfully any flickers of foresight were easily extinguished by the excitement of creating a really striking boundary between the house and garden. Extravagant perhaps, but delightful too.

*Steve and Dennis begin work on the pond.*

# Passive cooling

It's four months since work began and life on our tiny Clapham building site is full of the joys of mid-winter: shivering in the shed, lingering in the caff and bunking off as soon as the light begins to fail.

Fortunately this behaviour is only typical of the owner and not of site regulars Steve, George and Dennis, who are out in all conditions, grimly determined to finish our interminable ground works. Although the slab was laid before Christmas and now sits proudly on its nine-metre legs, there remains the small matter of our pond.

The main living space on the ground floor of Tree House has been designed as an inside-outside unity of three different ecological niches. Inside, humans and small furry mammals will enjoy their fossil-fuel-free comforts, gazing through a wall of glass to the small but richly planted courtyard garden, designed to attract bees and butterflies. Between the two, along the entire width of the house, a formal pond will provide a striking transition, itself home to a new community of many-legged beings.

When I dreamt up this little ecological fantasy eighteen months ago and scribbled it down for our ever-patient architect, I had no idea that I was kickstarting a major piece of engineering.

Because the pond sits immediately next to the house, it has to be integrated into the slab, and because it is relatively large, it has to be made of reinforced concrete and sit on piles of its own. On Thursday, after several days hard labour preparing the shuttering and bending the (recycled) steel rods into place, a truck laden with (recycled aggregate) concrete arrived to complete the job.

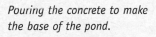
*Pouring the concrete to make the base of the pond.*

All this effort and expense will do more than improve the view and the local biodiversity. It will also help to keep us cool – and keep power-hungry air-conditioners out of harm's way. As glass walls can rapidly turn a house into a greenhouse, we are pulling every trick in the sustainable design book to stay comfortable.

The design of the living/garden space is inspired by the classical Roman peristyle garden with its close integration of inside and outside and careful use of water and plants to keep the space cool. The citizens of Pompeii did not incorporate formal pools into their houses merely because they looked pretty but because the evaporation of water draws heat out of the air – 600 calories per gram of water. Fountains are particularly good at encouraging evaporation, so it is no surprise that they are features of courtyard gardens from the Alhambra to the Moghul palaces of north India.

Trees and other plants perform a remarkably similar role: it's not just their shade that helps to keep us cool but the transpiration of water from their leaves. Trees sweat rather a lot of water, much more than they use for photosynthesis, nutrient transport and their own cooling. Evaporation from leaves in the canopy draws water up from the roots through minute columns of water that are kept in tension all the way up the tree. As so much water is evaporated just to keep this engine going, rainforests are able to produce their own cloud cover and good urban

*The first pond life is feline.*

38

planners are able to prevent their towns overheating. With water and plants in the heart of our living space, Tree House will keep us cool in much the same way that urban trees keep whole cities cool.

If your house suffers from summer overheating, think first about shading your windows (on the outside) and encouraging ventilation across the building or up inside it. If, however, you are still tempted to phone the air-conditioner company, ask about evaporative coolers. If the air is reasonably dry, you will get far more energy-efficient cooling. Better still, try a visit to your local plant centre. It might make the difference between living with an ugly, noisy box and the delight of a lush interior garden.

*design detail* | Victorian hand pump

The cooling potential of our pond has a downside: as the water evaporates, the pond level inevitably falls. As we do not want to use mains water to refill our pond, this loss could become a significant problem.

Our solution is to let the pond overflow into our deeper rainwater cistern when the rain falls, then pump the rainwater back into the pond in dry periods. Our hand pump is a beautiful Victorian original made from cast iron and brass. It may only be a hand pump but its robustness and durability express the confident values of the age in which it was made.

## Dennis

Dennis joined the team when the weather turned nasty in December and left just when the spring warmth was returning. The conditions couldn't have been worse for preparing steel rods and pouring concrete but Dennis never seemed to mind. Most of Dennis's work is now hidden beneath the house but the frogs regularly croak their appreciation

*resources*

### Design detail
Reclaimed hand pump from Holloways of Ludlow (01584 876207, 020 7602 5757, www.hollowaysofludlow.com).

### Publications
*Reducing Overheating – a designer's guide* (Energy Efficiency Best Practice in Housing CE129). Free to download or order from the Energy Saving Trust (0800 915 7722, www.est.org.uk).

*The Wildlife Pond Handbook* (L Bardsley and The Wildlife Trusts, New Holland 2003). A detailed guide to garden pond design with wildlife in mind.

# 02 FEB 05

By the beginning of February, our slab was well and truly finished. All we needed now was the timber to build the house on top of it. As there was no sign of it, I decided to take matters into my own hands: if the timber could not come to us, I would go to the timber.

One of the problems we faced at the time was finding a sustainable source for the engineered timber components that our architect had specified. James Jones and Sons of Forres came to our rescue and I had the genuine pleasure of being shown round their factory in the cold, wet twilight of a Scottish winter. Back on site we could only anticipate the spaces that the timber frame would create.

*Ford imagines the front door of timber Tree House.*

# Sourcing timber

The clouds hung low over our Clapham building site this week. Someone has been fly-tipping next to our skip, spilling rubble across the street. Site foreman Steve has been doing his best to clear it up, but it's a miserable business. Let's hope the perpetrators choke on their own cement.

I got out of the way, taking the Caledonian Sleeper to the north of Scotland to visit a factory that produces a truly remarkable construction material. It's similar to fibreglass in its design, a composite of tightly woven fibres and resin that gives great strength while remaining light and flexible. It's also tough – absorbing serious knocks without shattering – but is stiff enough to hold up considerable loads without buckling. To cap it all, it's made from renewable resources using 100% renewable energy.

Bizarrely, many people remain doubtful about the material's ecological credentials. I am not among them – after all, if Tree House is to express the beauty and integrity of the tree that towers over it, we can only build with the wonder material that is wood.

Anxieties about wood run deep. Logger's trucks and wrecked rainforests have become iconic images of ecological devastation and the most high-profile eco-warriors in Britain have been dendrites, tree dwellers, facing down the forces of the state from the canopy of ancient woodlands. Trees are venerated across the world and we rarely take pleasure in their loss.

*Brian Roberton and the timber I-beams which he was the first to manufacture in Britain.*

Such anxieties are well placed. It is scandalous, for example, that we continue to import tonnes of illegally felled timber from the rainforests of Indonesia and Brazil. Nearer to home, logging in the old growth forests of Finland has been criticised for disrupting the reindeer grazing grounds of the Saami people.

Happily, thanks to the work of the Forest Stewardship Council (FSC) and others, the ecological contradictions of buying wood can be overcome. If you buy wood that is FSC-certified, it will have a documented and independently audited 'chain of custody' tracing its journey from a forest that is managed in a manner that is 'environmentally appropriate, socially beneficial and economically viable'.

My visit to the Moray Firth retraced part of the chain of custody of the wood that will provide the structural skeleton for Tree House. The manufacturer, James Jones and Sons, has just gained FSC certification for its specially engineered timber products. For all its qualities, wood does present the odd problem, above all its tendency to shrink, twist and move – witness all those Elizabethan houses, delightfully free from right angles. Our 'I-beams', made from compressed timber particles, will never shift. I-beams also provide excellent thermal insulation in walls, use relatively little timber, and are easy to handle on site.

FSC-certified products, from sheds to salad bowls, are now widely available. The big DIY stores stock FSC wood, so it's not hard to find – check out the product selector on the FSC website. If you're buying from a timber yard, ask for FSC and your invoice should have a number on it with 'COC' (chain of custody) in the middle. If you want to know exactly where the wood came from, ask to see the documentation. If you're in B&Q or Homebase, you won't get this level of detail but their FSC timber should be clearly packaged.

An alternative approach to sourcing sustainable timber is to visit your local woodland and meet the trees. If you know the woodsman and can see the wood, you should be able to form your own judgement about the quality of its environmental management. A 'constant cover' forest where small areas are felled at any one time, keeping the forest as a whole intact, is more sustainable than clear-felling which wipes out both the forest and the local biodiversity in one go.

I got the sleeper back from Inverness on the evening of the 25th, Burns night. As we sped through the Grampians, I imagined a hundred thousand haggis being slain around me, bursting from their shiny skins. As my brother Hugh will tell you, I'm about as Scottish as he is green. One day, perhaps, I will return to my roots – and he will change his lightbulbs.

*design detail*

## Bathroom cabinet

Our bathroom cabinet was made by Simon Burrows from a yew tree that was felled in the churchyard of the Lincolnshire village where he lives. When I went to collect the cabinet, Simon was able to show me not only the seasoned yew in his workshop but also the stump of the tree and the perfect English setting where the timber for the cabinet grew and flourished.

Yew is a richly coloured and patterned wood. Simon's cabinet makes the most of these qualities, letting the beauty and texture of the tree shine out from within a clear, elegant frame.

## *resources*

The FSC engineered timber I-beams for Tree House were manufactured by James Jones and Sons (01309 671111, www.jji-joists.com) and supplied through Ridgeons Forest Products of Sudbury (01787 881777, www.ridgeons.com).

The Forest Stewardship Council in the UK provides information on sources of sustainable timber and timber products (01686 413916 www.fsc-uk.org).

In Europe, the PEFC label (Programme for the Endorsement of Forest Certification) indicates that the timber has been sourced under a national accreditation scheme. It is not yet as robust as the FSC label but is a good second-best (01829 770438 www.pefc.co.uk).

### Design detail
Yew and walnut cabinet by Simon Burrows (simonburrows@onetel.net).

### Publication
*The Good Wood Guide* (Friends of the Earth/Flora and Fauna International 2002). Advice on buying sustainable timber including an A to Z of woods and the threats they face.

# *09 FEB 05*

**This cheerful diary entry disguised a rather less jolly reality on site. Our timber frame was going to be late – very late. It took ages for the final specification to be agreed, partly because so many people seemed to be involved: the contractor, architect, engineer, manufacturer and supplier. It didn't help that the design was far from straightforward and required several rethinks before everyone was confident that the thing would stand up. In the meantime, we got on with some rather less pressing tasks.**

## Reclamation

This week I finally faced up to a job that I have been shirking for months. We have a shed-load of Burmese teak in need of serious TLC: dozens of bags of grubby reclaimed parquet, crammed in tightly from floor to ceiling. When I bought the lot last year from Lassco, London's biggest reclamation warehouse, I had a vision of the upper floors of Tree House glowing with the rich, warm hues that teak is famous for. Cleaning up 750 square feet of the stuff might be a bit of a challenge, but hey – I had plenty of time. Six months later, I decided it was time to invite our friends Nick and Hussein round to lunch, promising an interesting encounter with our building project.

In fact it was a good purchase, for each tile requires fairly brief attention. It's just that there are so

*Earlier – the bagged-up parquet awaits collection in Lassco's caverns.*

45

*Teak strip for the bathroom floor is also part of the order.*

many of them. Life gets difficult restoring parquet when the original installers have been too liberal with their bitumen, resulting in the joins between the tiles getting clogged up with the black gunk. The tops will be sanded down after the floor is relaid, so the key task is cleaning the slots and grooves that connect the tiles together. Happily, our grooves are only caked with the accumulated fag ash and skin peelings of several generations of Chelsea's water engineers.

Teak is a remarkable hardwood. As well as being attractive, it is very strong and naturally durable and so is a popular choice for tough assignments such as boats, bridges and British garden furniture. If you ever had a chemistry lesson on a wooden lab bench, it was almost certainly made from teak, as it is not harmed by acids and alkalis. The trees grow to 150 feet tall in the monsoon rainforests of south-east Asia, where traditionally they were sought out, ring-barked and left to die, then felled a year later before being dragged out of the forests by elephants. Unfortunately global demand for teak has led to unsustainable harvesting of teak but the sheer quantity that has been felled in the past, and the durability of the wood, means that there is a good supply through reclamation yards.

Reclamation is a crucial part of the recycling loop, keeping precious resources in use and eliminating the ecological costs of extracting new ones. It's particularly valuable when the resources in question come from threatened natural habitats such as rainforests. Reclamation yards also tend to be interesting places to visit. Like any other second-hand market, they are full of the promise of discovery: scramble past the piles of rusting Victorian radiators and the perfect piece of industrial archaeology lies awaiting your rescue and subsequent transformation into a chic garden ornament.

Unfortunately, reclamation is not without its risks. After all, there's nothing ecological about buying stolen goods. Within the industry, Salvo has developed a code of practice to limit these risks, but as yet there is no assessment of those who sign up. As well as its excellent online directory of salvage yards, Salvo also alerts buyers to architectural items that have been reported stolen. Lassco argues that their business is sustained only by reputation, so they cannot afford to take chances with goods of dodgy provenance. At the very least, always ask.

Although second-hand materials are often robust and hardwearing, their reuse is not always straightforward. Always consider exactly what you will need to do to a salvaged resource to make it function effectively within your home.

To be honest, Nick and Hussein didn't make much impact on our bags of parquet. Perhaps it was the wine; perhaps they were just too well dressed. As I'm not going to pull off Tom Sawyer's fence-painting trick with our easily distracted friends, I shall wait until the weather warms up, shut myself away, then dream of dancing the tango on a herringbone floor as the bags spill open around me.

*The reclaimed teak bathroom floor – hard work, but worth the effort.*

*design detail*

## Bathroom sink

The guys on site were amazed at the things I would turn up with to install. They were building a slick contemporary house yet all sorts of old junk seemed to be going into it. I found this bathroom sink in a salvage yard hidden in a dank, dark wood just south of Berwick-upon-Tweed. I ventured there on a cold November day and carefully picked my way through a series of musty sheds in various states of disrepair.

One shed was full of old sanitary-ware, including heaps of bulbous old toilets and uncherished sinks. But amid all this gloom, this sink stood out. There was something about it that caught my eye: a graceful, unadorned form that seemed to capture the function and utility of a sink perfectly. A quick twenty quid and its long journey to the bathroom of Tree House had begun.

*resources*

The reclaimed teak floors in Tree House were supplied by Lassco (020 7394 2100, www.lassco.co.uk).

Salvo maintain a directory of salvage yards and an online marketplace (www.salvoweb.com).

BioRegional Reclaimed specialises in high-volume/low-value reclaimed materials, such as structural steel, that most salvage yards ignore (www.bioregional-reclaimed.com).

### Design detail
Sink from Woodside Reclamation (01289 302 658, www.redbaths.co.uk).

## 16 FEB 05

Although the delivery date for our timber frame felt like an ever-receding point in the future, work never stopped on site. There was still plenty to do, not least digging the drains. This was a tricky business because we had to dig a channel under the roots of the tree to lay the waste pipe. As the job was done by hand some major roots were saved but the damage to the finer roots near the surface of the earth was substantial. Happily, the tree flourished through the following summer and will hopefully last for many more.

*Hand-digging the drainage trench is laborious, but major tree roots are saved.*

*Steve inspects the foundations of our manhole.*

## Drainage

Although I am desperate for our build to start going upwards, this week Steve was again going in the other direction. Directly under our tree, at the front of the site, he was putting the surprisingly complex finishing touches to the manhole, our private entrance to that least cherished domestic environment, the drain.

The tree does not have a drainpipe but deals with its wastes either by packing them into its trunk or by shedding leaves for the creatures below to consume and transform into new nutrients. Although we are following this example by designing recycling into the heart of Tree House, we're not pursuing the principle to its logical conclusion. We will not, I'm afraid, be installing a compost toilet.

Compost toilets come in many shapes and sizes, everything from 'bucket loos' to special digesters in your basement. One way or another they turn your personal organic wastes into something you can sprinkle on the garden. They can be invaluable in areas where connection to a sewer is difficult or impossible, but they have an obvious downside that will always limit their ecological impact. Toilet flushing is the single biggest cause of water consumption in homes and compost toilets can reduce this by 100%, but because your waste stays with you such toilets will only ever be acceptable to around 1% of the population of Clapham. Our Ifö Cera ES4 toilets reduce water consumption by 50% but are 100% acceptable. It's obvious which technology has the greatest strategic water-saving potential.

Arguably, this mealy-mouthed arithmetic simply reflects a greater malaise. After all, high-density societies have survived with wholly localised nutrient recycling. The WC is one of the dubious pillars of modernity: just as Descartes' *Cogito, ergo sum* raised the status of the mind over the soul, so Thomas Crapper's 'flush and forget' has enabled millions to deny that they have a gut. We have become brains without bottoms and our many out-of-sight toxic waste streams may simply be an extension of our contemporary failure to confront our own shit.

*The drain takes shape. It will never look this tidy again.*

Perhaps. But I'm still not persuaded by the urban compost loo. Given that our end products are 90% water, the amount of compost a household can generate is fairly meagre. In cities at least, it makes more ecological sense to deal with this at a collective level, turning sewage sludge into a reliable renewable fuel or an agricultural fertiliser. Sewers are, after all, the foundation of modern public health.

If you want to cut the ecological costs of your drainage, there's plenty you can do without giving up your porcelain throne. Cut your water consumption with efficient toilets, high quality taps and efficient washing machines. Use recycled toilet paper, and toiletries and detergents that aren't stuffed with nasty chemicals and bleaches. If you're doing some DIY, make sure the paints and finishes you wash off your brushes are non-toxic and water-based. If you've got a compost heap, improve your personal recycling by adding your own liquid fertiliser (easier for some genders than others, I accept).

If, however, you are in a location or frame of mind where an alternative toilet might be attractive, you have lots of choices. Compost toilets are at one end of the spectrum, reed beds are at the other. This is definitely one area of specification where a little personal research goes a long way. You should never rush a decision about what to do with your personal gifts to the world.

*The finished manhole fits snugly amidst the slate-packed porous paving.*

design detail

## Gutters

Like the rest of Tree House, our gutters are bespoke and rather lovely. They were made on site by Hughie the metalworker and carefully integrated into our very high performance roof.

The gutters are made from stainless steel. Although steel is not exactly an eco-product, stainless steel is exceptionally robust and durable and is therefore a good choice for a relatively small feature, such as a gutter, that will have to cope with severe conditions year after year.

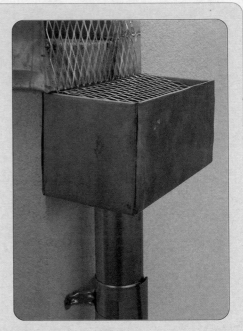

resources

The Ifö Cera ES4 toilets for Tree House were supplied by the Green Building Store (01484 854898, www.greenbuildingstore.co.uk).

For eco-friendly detergents and paints, try the Green Shop (01452 770629 www.greenshop.co.uk).

Elemental Solutions offers consultancy on natural water and sewage systems and water efficiency (01981 540728, www.elementalsolutions.co.uk).

### Publications

*Lifting the Lid* (P Harper and L Halestrap, Centre for Alternative Technology 1999). Alternatives to the mains-connected toilet.

*Sewage Solutions: Answering the Call of Nature* (N Grant, M Moodie and C Weedon, Centre for Alternative Technology 2005). A wider look at the sustainable ways of managing sewage.

## 23 FEB 05

February is rarely a joyful month and I was finding it difficult to keep my spirits up when there was so little progress on site. Nonetheless I remained philosophical: this diary entry was a patient meditation on the trials and tribulations of ambition, especially eco-ambition. At the time, I accepted that everyone was doing what they could and the delays were due to my insistence on the very best eco-specification. After many more delays, I was not quite so ready to give our contractor the benefit of the doubt. Nonetheless I stand by everything I wrote.

# The costs of building green

What does it cost to build green? This ought to be a straightforward question, but the straightforward answer – more! – is wrong some of the time and misleading all of the time. The ordinary builder's bottom line (and I'm not talking ill-fitting jeans here) does not take account of any environmental costs. Nor does it reflect

*Will admires his solar roof. If you want to make your own power, you have to pay the capital costs.*

the long-term value of durable materials, a healthy living environment and low running costs. Tree House may not be the cheapest self-build in Britain today, but once we're in we will have no fuel bills and we'll be able to enjoy a consistently comfortable, draught-free home that is filled with natural light and free from toxic pollutants.

Building professionals will also tell you that doing anything non-standard is unpredictable, risky and therefore more expensive. This argument is a little closer to the knuckle at the moment: we are currently paying a price for our ecological ambition in our embarrassingly absent super-structure. Our beautifully prepared slab, endlessly nipped and tucked by ever-patient foreman Steve, is beginning to feel like a launch pad for a rocket that left long ago, a lingering concrete impression of ever-receding dreams.

Last month I belatedly discovered that the company we originally approached to supply our timber frame had no evidence that it came from a sustainable source, despite the fact that their website was plastered with the Forestry Stewardship Council logo. Fortunately we were able to find a manufacturer, James Jones and Sons, and a supplier, Ridgeons Forest products, who were willing to pull out their FSC stops for us. But some re-engineering of our ultra-efficient walls was also required, adding to the delay.

Happily we have a contractor who does not see this as an opportunity to slap some extra costs on us but instead has jiggled his schedule, bringing forward some jobs to fill the gap, so that the costs of the delay will be limited to some hard-bitten fingernails. Our contract is not the result of competitive tendering but of a series of discussions led by our architect that began early in the design process, a 'negotiated contract' that is especially appropriate for an unusual build that is difficult to price. If you don't go out to tender you risk getting ripped off but the reward of a successful negotiation is a set of working relationships based on trust and a shared commitment to the building itself.

Enthusiasm for the building has proved to be a major benefit of our ambition, a payback for our determination to push beyond the fluffy green foothills of eco-design and into the more challenging landscape of energy self-sufficiency. This includes everyone from our very supportive neighbours to Chris the concrete man, eager to tell me about his recycled aggregate.

If you are embarking on a building project, however small, you will be under pressure from day one to stick to the familiar. But if you want professionals who are really going to take an interest, don't compromise. Whatever your design fantasies might be, eco or otherwise, let them rip. You can always sew them up neatly again if circumstances dictate.

Back on site, I remain nervous that one delay will lead to another, despite the best of all our intentions. Hopefully our engineered timber beams will arrive any day now and, amid a flurry of sawdust, Clapham's first carbon-neutral Apollo mission will finally leave the ground.

## design detail

### Door handles

Buy cheap, buy twice. This very distilled expression of the case for ecological design is highly applicable to any building component that moves, such as a door handle.

The handles and latches for the interior doors of Tree House were specified early on as high quality items. We sourced them from a Devon company that boasts a strong environmental policy and very rich design ideas. The leather trim is pure luxury, creating everyday moments of subliminal sensual pleasure.

## resources

To find a builder or other construction professional with green interests, begin with the membership of the Association of Environment Conscious Building (www.aecb.net). Member information is available through the online database or in *The Green Building Bible* (K Hall, Green Building Press 2005, www.newbuilder.co.uk).

### Design detail
Door handles by Turnstyle Designs (01271 325325, www.turnstyle-designs.com).

### Publication
*Using whole life costing as a basis for investments in energy efficiency: guidance* (Energy Efficiency Best Practice in Housing CE119). This publication shows how a systematic approach to counting whole life costs works out in favour of energy efficient purchases. Free to download or order from the Energy Saving Trust (0800 915 7722, www.est.org.uk).

# 09 MAR 05

The long wait for the structural timber was discouraging but it gave us time to put some effort into jobs that would otherwise have been neglected, particularly the considerable task of building our boundary walls and fences.

This task proved to be both labour-intensive and expensive. It would have been hard to schedule and impossible to finance at the penniless end of the project. Lesson learnt: don't put off jobs that will be crucial to your vision but easy to sacrifice when under pressure.

## Lime

The walls came tumbling down this week. Not, thankfully, the walls of our timber frame (which are yet to rise), but the old brick walls that form the boundary at the front of our site. It was a pity to lose them but they were leaning precariously

*Jonny scrapes the lime mortar off the bricks from the toppled wall.*

*The salvaged bricks ready for reuse.*

and various DIY bodges over the years had not done their structural integrity any favours. We want to rebuild them to last another century, so non-stop Jonny, our current site hero, knuckled down to the task of reclaiming them, brick by brick.

Most of the wall had been built with traditional lime mortar, a soft binder that is relatively easy to scrape off, but some sections had been made with the cheap and cheerful modern alternative – cement. These had set so hard that it was impossible to remove the cement without ruining the bricks in the process. The ubiquitous use of cement is a hidden eco-disaster because the life expectancy of any masonry building is far shorter than the solid materials it is constructed from: buildings fall apart before bricks crumble. Reusing bricks, time and again, would cut the huge environmental cost of the three billion new bricks baked in the UK every year.

Until the beginning of the twentieth century, lime mortar was used to build everything from houses to gothic cathedrals. Cement, invented in the early nineteenth century, eventually triumphed because of its structural rigidity, especially in the versatile form of concrete, a cement and rock mixture that can be poured to make any shape you desire. The death knell of English lime kilns came in the second world war when huge cement works were built to construct our cubist beach defences. After the war their output was diverted to reconstruction (in some unfortunate places only minor modifications to the beach model were required), rapidly wiping out the market for lime mortar.

But rigidity is not everything. Lime mortar may not bond as tightly as cement but this gives it a flexibility that enables walls to cope with pressure and movement without cracking. Unlike cement, lime mortar is porous and allows walls to breathe: any dampness inside a wall can pass through the mortar rather than being forced through the bricks, damaging them in the process. In the same way, a wall rendered with lime will let moisture escape whereas the same wall rendered with cement will encourage moisture build-up inside your house.

Lime may not be quite as brilliantly green as its small, round homonym, but it beats cement on all counts. Although lime and cement are both made from the same raw material – limestone – cement is fired at a much higher temperature, and although in both cases the chemical reaction involved gives off lots of carbon dioxide, lime mortar reabsorbs the gas when it is setting (or 'going off' as a brickie would say).

Lime mortar and lime render are widely used in restoration work but there's great scope for their revival in new buildings. If you have a wall or house to build, get advice from one of the specialist suppliers about which lime to use. But protect yourself when using lime as it burns.

## Front gates

We rebuilt our two-metre-high front walls to a height of only one metre because we were required by the planning consent to improve the visibility of any vehicles leaving the house. So we had some fences made to fill the gaps.

The gates and fences at the front of the house turn our arboreal vision into a rather outrageous gothic fantasy. Within a graceful curving frame, the steel rods of the fence warp and twist like the patterns in the bark of our tree. As you push the sycamore seed handles of the gates and enter, you are passing through the cambium of Tree House and into the magical interior.

*The brick walls are now one of the many attractive finishes for Tree House.*

## The brickies

The brickies seemed to spend most of their time drinking tea in their van. They were not, however, a bunch of idle shirkers. They simply couldn't do any bricklaying when the temperature was too low for the mortar to set.

Using lime mortar presented no problems other than the need for care when mixing the stuff. The team told me that they had been trained using lime precisely because you can knock everything down and start again with relative ease.

The front walls of Tree House are both beautifully patinated, thanks to the rich colours of the reclaimed stocks, and beautifully pointed, thanks to the skills of the brickies.

Specialist lime suppliers include:

Lime Technology (0845 603 1143, www.limetechnology.co.uk).

Old House Store (0118 969 7711, www.oldhousestore.co.uk).

The Cornish Lime Company (01208 79779, www.cornishlime.co.uk).

You can attend courses in using lime at the Lime Centre (01962 713636, www.thelimecentre.co.uk).

### Design detail
Metal design by Jonnie Rowlandson (020 7720 6428, www.argonautdesign.co.uk).

### Publications
*Building with Lime* (S Holmes and M Wingate, ITDG 1999). A comprehensive guide to using lime in both new and historical buildings.

*Lime in Building: A practical guide* (J Schofield, Black Dog Press 1997). Concise advice.

# *16 MAR 05*

**At the beginning of March the temperature dropped even further and the site was soon covered in snow. I did wonder if we should just pack up for a couple of weeks and have a holiday but there were always jobs for our little two-man team to get on with. Connecting to the sewer was an important task that offered a degree of shelter from the elements even if, in every other respect, conditions were less than ideal. Uncle Steve sent his nephew down the drain to sort it out.**

## Ultra-low-heat houses

So is this it? Is this my heart of darkness? Just when the pace of our ambitious Clapham self-build was finally picking up, the gods have intervened, rendering our site a frozen wasteland where little, if anything, can move forward. The brickies have been sent home because it's too cold for their mortar, the timber frame has been put back for yet another week and site stalwarts Steve and George are officially fed up to their back teeth. George spent a day last week twenty feet under at the bottom of our neighbour's manhole, drilling through to connect to our drain. Most of the residents of the seven houses which supply the manhole were out at the time, but it was a pretty good image of existential desolation: wading through human effluent in a dank dark hole while the freezing sleet beat down from above. At least he got to wear a fetching yellow outfit for the occasion.

*Our hero at work.*

But even snow clouds have silver linings. The sudden winter has provided me with a chilling incentive to return to my long-neglected energy calculations and consider how well Tree House will cope with similar conditions in a year's time.

Most of the energy we burn in our homes goes on space heating, so there's nothing like a cold snap to bring a warm glow to the shareholder value of energy companies. Yet this needn't be the case. There's a common misunderstanding that when it gets cold outside, our houses inevitably get cold too unless we turn the heating on. But the problem isn't really the cold at all. We only need heating because our houses fail to prevent the heat that's already in them from escaping, so although a drop in outside temperature will speed up the heat loss, this loss is principally due to the construction of the building.

There's no reason why a house shouldn't remain comfortable throughout the year without any active heating, as long as the heat losses never exceed the 'passive' heat gains from the sun, the occupants and their cookers and appliances. In practice, this is very difficult to achieve, although at BedZED in South London they have come close by trapping the sun's energy in their huge concrete walls. Similarly in Germany the Passive House movement has successfully promoted a radical design specification that involves minimal active heating.

We will do pretty well: on a freezing morning Tree House will lose heat at a rate of 2200W, a loss that we could replenish using a two bar electric fire for the whole house, whereas a 1930s house of the same proportions will throw heat away nine times faster, requiring a hard-working boiler to keep the temperature up. We don't want to ditch heating systems altogether, because we'll always need hot water, so our approach is to combine a very high performance building (exceptional insulation, high quality windows and carefully controlled ventilation) with our own renewable energy generation.

If my sums are right, in a year's time we will be able watch the snow settle on the boughs of our tree from a warm, fossil fuel-free Tree House. I only hope that in fifty years' time our rapacious oil-driven global economy will have avoided its own heart of darkness and we will still be able to potter down to Brixton market without being chased by daily Arctic gales.

*When the clouds gather, this house stands proud. A very low-energy house must be robust in its adaptation to climate.*

**design detail** | Slate floor

As we wanted to extend our ground-floor living space out of the house, across our pond and into our courtyard garden, we incorporated a rather extravagant amount of glazing into the west elevation of the room. The computer modelling of the building demonstrated that all this glazing would result in regular summer overheating.

We tackled this problem by installing substantial external shading and by tiling the floor with Kirkstone slate. The stone absorbs the heat of the sun and helps to moderate the temperature in the room over the afternoon and evening.

We used Kirkstone Silver Green slate for both the floor and the paving in the courtyard garden. It is extraordinarily beautiful, full of the patterns of millions of years of geological upheaval. Every tile is different with its own rich seam of primordial turmoil.

**resources**

Extensive advice on improving the energy efficiency of your home is available free from the Energy Saving Trust (0845 727 7200, www.est.org.uk).

### Design detail
Slate floor by Kirkstone (01539 433296, www.kirkstone.com).

### Publications
CEPHEUS. *Living comfort without heating* (H Krapmeier & E Drössler, Springer-Verlag 2001). This European study of very low-heat homes shows just how far heat demand can be reduced without sacrificing quality of life. See also www.cepheus.at.

*Review of ultra-low-energy homes* (D Olivier & J Willoughby, Energy Efficiency Best Practice Programme GIR 38). A dated but useful guide to what has been achieved in the UK and abroad. Free to download from the Energy Saving Trust Best Practice in Housing archive.

Beddington Zero (fossil) Energy Development: www.bioregional.co.uk.

The Passive House Institute: www.passivehouse.com.

# 23 MAR 05

**This diary entry describes the arrival of a small but crucial consignment of steel. Unfortunately, however, the steel didn't fit and had to be sent back for refabrication. This became our next big delay, even before the timber for the bulk of the structure had arrived.**

**We could have built Tree House without steel but we would have had to sacrifice our big southern pitched roof and our integrated garden-living room. We wanted the former to achieve our energy self-sufficient goal; we wanted the latter to achieve the inside-outside space we had always dreamed of. Despite all the problems that the steel brought us, we are glad not to have given up on these ambitions.**

## Steel and scrap

Tree House, a house that works like a tree, is a personal ecological fantasy slowly being made real in a quiet corner of Clapham, south London. It is driven by my belief that ecological design can be a 'win-win', delivering great results for us as well as for the planet, an idea that I hope is no longer fantastical in itself. But we want more than this: we want to create sensational living spaces in which

*The steel arrives for a short stay.*

*Jonnie searches the scrapyards for raw materials.*

inside and outside merge, where the form, light, colour and textures of the organic world surround us. We want Tree House to be a stylish contemporary home, but we also want it to be magical.

In practice, ecological principles and our design ambitions have not perfectly converged. To be blunt, the occasional compromise has been necessary. This week our long-awaited super-structure started arriving on site but the first components to be delivered were not made of Douglas fir nor engineered timber but that most un-green of building materials, steel. Although you can build six-storey buildings in wood, steel does come in handy when you're trying to be clever with your interiors and our open plan living space, extending through a glass wall into our courtyard garden, would not be possible without a steel joist to keep the rest of the house up.

The metal-smelting industry is second only to the chemicals industry in the toxic emissions it produces and iron and steel production has the worst record for water pollution in the UK. The raw materials are mined, transported huge distances and then smelted in blast furnaces, consuming huge amounts of energy at every step. And the problems aren't over when the steel gets on-site: such a good conductor can suck the heat out of a building unless every post and beam is highly insulated.

Nasty stuff. But as I wouldn't dream of leaving you with anything other than a warm eco-glow, let me persuade you that in the right hands even steel can be an eco-product.

*The flames of the fire fence, burning in March.*

Enter Jonnie Rowlandson, metal-worker extraordinaire, based up the road in the Clapham North Arts Centre. Over the last month, Jonnie has been scouring the scrap merchants of south London for sheets of rusting steel, old copper hot-water cylinders and fragments of discarded brass. These salvaged materials have excellent eco-credentials for whenever you buy something salvaged or second-hand you are keeping valuable resources in use and avoiding the extraction and manufacture of new resources.

With this gritty palette, Jonnie has made us the ultimate garden fence. Because we will have no fireplace in our ultra-low-energy house, our garden will be our hearth and the back fence will be the focus of the hearth. Jonnie has risen to this metaphor and created a fiery, organic, richly-textured work of art, alive with the dense, complex colours of rusted and stressed metals. Like any fire, it will change over the day and over the seasons, a constant source of fascination and inspiration. In the middle of winter when the garden is quiet, this fire will blaze.

So although a well-hidden and thoroughly un-eco steel joist is critical to the engineering of our open-plan living space, it is rusty, salvaged steel that will transform our experience within it. Thanks to our local alchemist, the first expression of our magical ambitions has been successfully forged from the basest of metals.

## design detail

### Danish copper table

I have always been a fan of copper. The Egyptian hieroglyph for copper is the Ankh sign (☥), which also means eternal life. The modern name derives from the island where it was mined in Roman times: Cyprus.

The warm colour and rich patina of copper are a core element in the palette of Tree

House, alongside teak, beech, Kirkstone slate, red linoleum and foliage green. Other than the copper in our wires and pipes, all the copper in the house is either salvaged or second-hand. This vintage Danish table has a copper frame and a surface made from beautiful red ceramic tiles.

*The installation of the fence proves to be a tricky job.*

## Jonnie Rowlandson

We needed a metal-worker for our fences and front gate. But where, I wondered, would we find one who had the imagination to respond to our guiding metaphor, the tree?

After a couple of false starts, I tracked Jonnie down in his workshop in Clapham. I rapidly found out that not only was he a skilled metal-worker, he was also especially interested in using organic forms as design inspiration. Unbelievably he had already worked with sycamore seeds. Here was someone who did not look at us askance when we babbled on about trees but instead rose to the brief with enthusiasm.

*resources*

The fences and gates for Tree House were designed and built by Jonnie Rowlandson (020 7720 6428, www.argonautdesign.co.uk).

BioRegional Reclaimed specialises in high-volume/low-value reclaimed materials, such as structural steel, that most salvage yards ignore (www.bioregional-reclaimed.com).

The 350 members of the British Metals Recycling Association recycle 10 million tonnes of metals every year (www.recyclemetals.org).

### Design detail
Danish table from Inside (www.insideoriginals.com).

# 30 MAR 05

Six months in, the component parts of our timber frame finally arrived. This should have been a moment to celebrate, but the sight of piles of damp timber covered with blue tarpaulins made it abundantly clear how much there was still to do.

It seemed rather unlikely that this could all be turned into a beautiful building in a matter of months, so an invitation to visit Steve's very own hand-built house provided the perfect opportunity to focus on the rewards awaiting us at the end of the journey.

# Timber-frame buildings

After our instant winter, the instant spring. Only two weeks ago the streets were covered in snow and our building site was icily inactive, but today Brixton is crammed with a heaving crowd and the mood of our little eco-building team is significantly warmer.

Steve Archbutt, our site foreman and principal builder, has more reasons to be cheerful than the rise in temperature, notably the arrival on site of the first components of our timber frame. Although he has spent most of the last five months digging holes, shifting dirt and pouring concrete, Steve is first and foremost a carpenter, with a carpenter's delight in the smell of sawdust on a warm spring day. The delayed delivery of our frame has been frustrating, but at least we can now put

*Steve considers the task ahead.*

it up in a kinder and perhaps more appropriate season. Tree House will rise as the towering tree next to it unfurls its buds and begins its own new year of growth.

If 'timber frame' makes you think of magnificent oak structures or the crooked charm of Elizabethan wattle and daub houses, think again. People do still build houses out of massive beams of green oak, but most timber houses are made using a softwood skeleton, hidden away. Our timber frame will be disguised by plasterboard on the inside and protective cladding on the outside. Although wood will be a defining feature of Tree House, the only exposed structural timbers will be the spreading branches of Douglas fir that hold up our solar roof.

*Steve, Diane and Sid on the balcony of the house that Steve built.*

As the structure can be hidden, anyone can enjoy the ecological benefits of a timber frame regardless of their personal domestic aesthetic. Timber is of course a top-notch renewable material made from carbon dioxide with solar power. Timber structures tend to be flexible and adaptable, so fewer resources are needed to keep them viable over a lifetime of occupant desires. Finally, timber-frame houses are usually very well insulated. In a brick-and-block house, insulation is stuffed in the cavity between the blocks and the bricks, whereas in a timber-frame house almost the entire wall can be made of insulation, as only the posts in the wall bear the structural load. Our posts are engineered 'I-beams', with very thin middle sections, enabling us to cram in even more insulation.

I'm delighted that Steve is building Tree House as he has an impressive twenty-year track record of putting up unusual timber buildings. This includes his own house, one of the celebrated Walter Segal designs in Lewisham, south London. Segal, whom Steve got to know in the early eighties, developed a modern approach to 'post and beam' timber building that is flexible, relatively cheap and ideal for self-builders. The Walter Segal Self-Build Trust promotes his ideas today and many attractive and award-winning schemes continue to be built using the Segal method.

After building so many houses for other people, Steve is itching to build again for his own family. So if you live in south London and happen to have a spare patch of land with its own road frontage, do get in touch. You never know, you could be Steve's very next reason to be cheerful.

*On this plan of the first floor, the vertical I-beams can be seen positioned at regular intervals in the walls.*

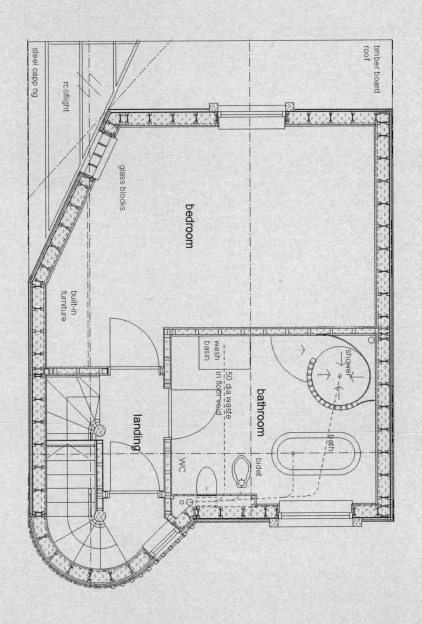

steel capping

rooflight

glass blocks

built-in
furniture

timber board
roof

bedroom

landing

wash
basin

50 dia waste
in floor void

WC

bathroom

bidet

bath

shower

## Stair tower cladding

Most of our exterior walls are covered in a bright white render board but for the all important stair tower, the trunk of Tree House, there was really no choice but to use timber. The long vertical slats are western red cedar, the smaller posts that rise between them and become part of the balcony balustrade are Douglas fir. Both these woods are highly durable though their colour will fade with exposure to the sun, turning a silvery grey.

Cedar has become a popular material for cladding in recent years. Just as many timber houses disappear behind a cladding of brick or render, so today many masonry houses now sport attractive timber façades.

*resources*

The FSC engineered timber I-beams for Tree House were manufactured by James Jones and Sons (01309 671111, www.jji-joists.com) and supplied through Ridgeons Forest Products of Sudbury (01787 881777, www.ridgeons.com).

For information about contemporary use of timber in construction contact the UK Timber Frame Association (01259 272140, www.timber-frame.org), the Timber Research and Development Association (www.trada.co.uk) and Wood for Good (0800 279 0016, www.woodforgood.com).

Many community self-build projects use timber construction. Contact the Community Self-build Agency (020 7415 7092, www.communityselfbuildagency.org) and The Walter Segal Self-Build Trust (www.segalselfbuild.co.uk).

### Publications

*Architecture in Wood: A world history* (W Pryce, Thames and Hudson 2005). A photographic survey of timber architecture.

*Out of the Woods: Ecological designs for timber-frame housing* (P Borer and C Harris, Centre for Alternative Technology 1999). A guide to the Segal method of timber-frame building.

*Timber Building in Britain* (R W Brunskill, Cassell 2004). The history and methods of British timber-frame construction.

## 06 APR 05

Ah, the naïve optimism of spring. In April a completion date in September still seemed entirely feasible so I was determined to have a garden ready to enjoy when we moved in. Planting a garden on a building site is barking mad but that didn't put me off. Over the summer I was able to tend our little courtyard garden as it flourished and subsided in the midst of all the building works.

While I planned the garden, Steve was counting his timbers. Before the vertical wall studs could be erected, a 'sole plate' made from oak had to be installed over the concrete-block base of the walls. This was the layer of transition from the harsh concrete of the foundations to the more forgiving, flexible realm of the timber frame.

# Organic gardening

Spring is in the air and our sap is rising: not only is Tree House finally reaching up to the light but the buds of our great multi-stemmed sycamore are looking distinctly brighter despite six months of abuse. We have done what we can to protect the tree but drain-digging, muck-shifting and scaffold-erecting have all had an impact. No major roots have been cut but many of the critical fine roots in the top 600mm of ground have been lost. Fortunately sycamores are tough trees with unusually deep tap roots, so I am hopeful that tree and house will continue to flourish together this year.

The tree is a central part of our biodiversity strategy, providing high-rise habitation for some very diverse fauna (including, I fear, rather too many aphids) and shelter for the bird and ladybird sanctuary in our front garden

*Steve checks the level of the oak sole plate.*

(we're hoping that attention to the ladies will keep the aphids in check). At the other end of the site, our small back garden will be a little organic haven attracting some serious pond life and plenty of bees and butterflies. The space is currently occupied by the site office (aka Steve's shed) but this has not deterred me from my annual Easter ritual of over-enthusiastic shopping trips to rural nurseries.

A mature organic garden is a perfectly functioning mini-ecosystem in which soil, plants and creatures of all kinds maintain a rich and self-sufficient 'circle of life'. The foundation of any organic garden is the soil. Maintaining the soil requires the regular addition of organic matter, either dug in or applied as a mulch to be drawn down by worms (the latter strategy is better in the long term for the structure of the soil).

A really healthy organic garden will be less likely to suffer from infestation than a chemically controlled garden. Pests that do threaten can be tackled with biological controls or by selecting plants that will not be attractive to them. One advantage of our constrained urban garden is that it will be small enough to deal with the perennial snail problem by hand.

Although many modern flower cultivars and introductions from abroad are only of interest to humans, it isn't difficult to incorporate choices that bees, birds and butterflies will also enjoy. We're hoping to be in Tree House by September, so I am planning a late summer fire for our garden-hearth. The main attraction for bees will be a mass of scorched heleniums – bees prefer clumps to scattered flowers – including my favourite, 'Waldtraut', and the towering 'Indian Summer'. Further forward I'm offering them coal-black scabious and the glowing embers of *Geum coccineum*. Butterflies will also enjoy the scabious but their attention is likely to focus on the smoky plumes of *Buddleia davidii* 'Black Knight'.

*George and Con prepare the ground that will eventually become the courtyard garden.*

The combined effect of modern industrial farming and Ground Force has been a shift in the heartland of biodiversity towards our gardens, not least those in the centre of our cities. The more we can improve our garden habitats and reduce the toxins we pour over our urban environment, the more our urban flora and fauna will flourish – including of course the two-legged kind.

## design detail

### Frog hideouts

Female mosquitoes lay their eggs on the surface of sheltered water. Our substantial pond is therefore an ideal breeding ground for this unwelcome domestic pest. The best way to deal with the problem is biological control, above all, frogs. To encourage the frog and tadpole population, the pond has shallow edges with a raked bank to the deeper centre and cool hiding places among the rocks. Frogs spend most of their time lurking on the land, so it's important to give them damp hideouts and easy access in and out of the pond.

## resources

The leading UK organisation promoting organic gardening is the Henry Doubleday Research Association (HDRA: 024 7630 3517, www.gardenorganic.org.uk).

A month-by-month series of Gardening Tips is available from the Pesticides Action Network (www.pan-uk.org).

Green Gardener offers organic birdseed and a variety of biological controls for garden pests including slugs (01394 420087, www.greengardener.co.uk).

### Publications
*The HDRA Encyclopedia of Organic Gardening* (Dorling Kindersley 2005) is a very accessible resource. Other smaller publications from the HDRA include *Pests and How to Control Them*, *Controlling Weeds Without Chemicals* and *The Small Ecological Garden*.

*Organic Gardening* (P Pears and S Strickland, Royal Horticultural Society 1999). A thorough guide.

# 13 APR 05

Over six months after the build began, our first walls went up. After such a long wait, the moment was sweet.

The engineered I-beams that we used for the wall studs and the floor joists all had to be cut to size on site. A prefabricated frame would have been quicker but our design was too quirky and bespoke for this to be an option. Although off-site prefabrication is promoted as a means of improving the quality of construction, especially for mass housing, you can't beat a chippie with an eye for detail and a commitment to high quality results. As Steve was responsible for the job on site from start to finish, he knew that accuracy at this stage would pay off later.

# Future-proofing

In a famous episode of the Simpsons, Homer falls through the back of a hall cupboard, Narnia-like, into an undiscovered world beyond. As family and friends gather outside, Frink reveals the shocking truth of the land beyond the coat-racks. He draws two overlapping squares on the wall and then, to gasps from his audience, joins their respective corners. Homer has fallen out of cartoon flatland and into the third dimension!

*A line of vertical I-beams defines the very first wall of Tree House.*

I shan't claim that Homer is a hero of mine but I've had his cupboard in mind for some time. Because the ground works of Tree House took so long, I became overly focussed on the low horizons of our concrete slab and lost sight of the pregnant nothingness above it. Now that the walls are finally going up, the feeling of smallness and constraint engendered by the building's footprint has been transformed into wonder at the size and possibilities of three-dimensional space.

*The new dimension brings a smile to Will's face.*

All great buildings achieve their impact primarily through their manipulation of space and transformation of the human experience within it, a subtle craft that can easily go wrong. The articulation of space is just as important in the domestic environment as it is in grand public buildings, but because the price of a house is so strongly determined by the number of bedrooms it boasts, indoor spaces are routinely chopped up into as many little boxes as possible. Consequently we all end up hacking our houses around endlessly to try to create the spaces we really want.

*The architect stands in the doorway of his creation.*

When we were designing Tree House it took time to think beyond the ways of living that our three-bedroom Victorian terrace had forced us into, but our reward will be interior spaces that are both exciting and ambiguous. In particular, the single large room at the top of Tree House, originally slated to be two bedrooms, will be a study, spare room, dance floor, yoga room and potting shed. If two bedrooms are required in the future, it will be easy to partition the space to create them.

The ecological importance of adaptable structures should not be underestimated. The more narrowly you define the interests of a building's users, the more likely it is that large quantities of resources, energy and patience will be expended over the life of the building in the endless struggle to reconfigure the space. In contrast, most traditional forms of shelter are strikingly adaptable. For example in Borneo, where I was born, the Iban of Sarawak are famous for their long-houses in which whole communities live together under one roof by sharing large communal spaces and using flexible partitions to create private rooms.

Although there is a rush to build millions of new homes for an increasing population and more one-person households, the population is expected to start falling after 2021. Factor in the complexities of changing work patterns and we can be confident that many different demands will be made on these spaces in the future.

*The adaptable second floor of Tree House will be used for many different purposes over the lifetime of the house.*

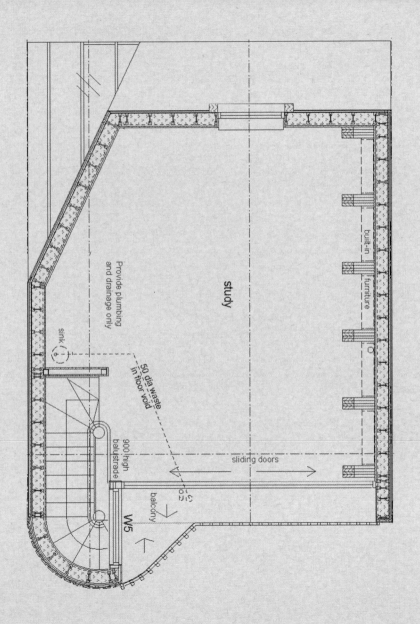

study

built-in furniture

Provide plumbing
and drainage only

sink

50 dia waste
in floor void

900 high
balustrade

sliding doors

balcony

W5

*design detail*

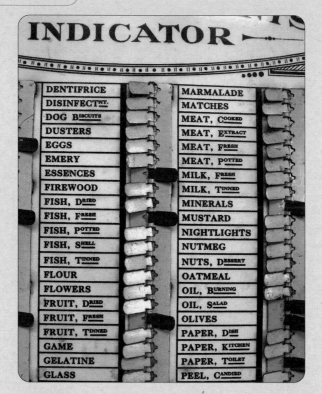

| INDICATOR | |
|---|---|
| DENTIFRICE | MARMALADE |
| DISINFECT<sup>NT.</sup> | MATCHES |
| DOG B<sup>ISCUITS</sup> | MEAT, C<sup>OOKED</sup> |
| DUSTERS | MEAT, E<sup>XTRACT</sup> |
| EGGS | MEAT, F<sup>RESH</sup> |
| EMERY | MEAT, P<sup>OTTED</sup> |
| ESSENCES | MILK, F<sup>RESH</sup> |
| FIREWOOD | MILK, T<sup>INNED</sup> |
| FISH, D<sup>RIED</sup> | MINERALS |
| FISH, F<sup>RESH</sup> | MUSTARD |
| FISH, P<sup>OTTED</sup> | NIGHTLIGHTS |
| FISH, S<sup>HELL</sup> | NUTMEG |
| FISH, T<sup>INNED</sup> | NUTS, D<sup>ESSERT</sup> |
| FLOUR | OATMEAL |
| FLOWERS | OIL, B<sup>URNING</sup> |
| FRUIT, D<sup>RIED</sup> | OIL, S<sup>ALAD</sup> |
| FRUIT, F<sup>RESH</sup> | OLIVES |
| FRUIT, T<sup>INNED</sup> | PAPER, D<sup>ISH</sup> |
| GAME | PAPER, K<sup>ITCHEN</sup> |
| GELATINE | PAPER, T<sup>OILET</sup> |
| GLASS | PEEL, C<sup>ANDIED</sup> |

## Household wants indicator

This Victorian gem has pride of place in the kitchen of Tree House: a permanent shopping list with little red tabs to flip backward and forward depending on whether an item is needed or not.

I inherited the indicator from my grandfather who had used it assiduously to keep track of his daily needs. Although he too was an old Victorian, even his needs changed over his lifetime and so he was forced to adapt the list. Remarkably (but typically for him) he managed to include new items by sticking labels over existing items that he no longer wanted in such a way that the additions maintained the precise alphabetical order of the list.

Although we have removed his labels we have not yet created our own twenty-first century adaptation. We're working on it though: feta instead of firewood, lager instead of lard, manure instead of mantles and vitamins instead of vim.

To create an interior space that will last and last without getting regularly ripped out and rebuilt, begin by reflecting on your own future and the many possible trajectories that this might take. If you are able-bodied, how will the space suit you if you relied on crutches or a wheel-chair? If you are single, what would a partner or children demand? If you ever need to take in lodgers, how easy would this be? If you are young, will you be happy being old and frail in the same place? Of course, this isn't just a matter of your changing needs but also of the people who move in after you.

'Future-proofing' can simply mean putting in services even if you're not sure you will need them. For example, we are running a water supply to the top of Tree House but we won't be bringing it out of the wall until some damp hobby of the future requires it. Similarly, if you are rewiring, run some cables up to your roof for the day when solar photovoltaic panels come within your means. To complement the creation of long-term flexible spaces, use robust hard-wearing materials. Fashions come and go but natural materials such as wood and stone always retain their elegance and quality.

Inauspiciously, the episode of the Simpsons described above was one of the 'Treehouse of Horror' Halloween specials. Having found himself in bulgy three dimensions, Homer faces disaster when he is sucked into a black hole. Thankfully, although our budget for Tree House is a little precarious, saying that we are being 'sucked into a black hole' can still be deemed a slight exaggeration.

## resources

Consumer information on domestic 'future-proofing' tends to focus on the installation of domestic networks that can carry data, telecommunications, audio and video. This usually involves star-wiring CAT5 cable to all the rooms in a house from a central hub. For more information see magazines such as *Digital Home*.

### Publication
*The Adaptable House: Designing homes for change* (A Friedman and S Grillo, McGraw-Hill Education 2002). A detailed guide describing the many ways in which adaptability can be designed into houses.

# *20 APR 05*

Much to my relief our tree burst into vigorous leaf with no signs of stress from the seven months of abuse at its base. Now that we were well and truly out of the ground, the risk of inflicting serious harm on the tree was much lower.

After our initial euphoria at the sudden appearance of wall-like structures, the pace slowed right down again while we waited for the steel to support the back of the house. Although this was frustrating, the more time I had to play with, the more I could obsess about the final details of the building.

Lighting was a particular obsession. I was determined to use low-energy lighting in a really creative way but I found it remarkably hard to settle on a final specification. Fortunately our architect treated our ever-emerging ideas with respect. He was no stranger to the unpredictable path of a good design process.

## Lighting

One of the advantages of timber-frame buildings is the speed with which they can be erected, demonstrated to great effect in the barn-raising scene in the film 'Witness' where an entire Amish community pulls together to build a huge barn over the course of a warm summer's day. Given the technical complexities of Tree House, I was not expecting quite such rapid results, but I did hope to enjoy at least a frisson of Harrison Ford's outsider awe.

*Joists awaiting their turn.*

Unfortunately, just as the erection of our own frame was finally gathering pace, foreman Steve decided to close down the site and go skiing. Skiing! Needless to say, he is under strict instructions to spend the entire week in the bar and avoid all contact with bone-breaking slopes. Happily, the week has not been lost as George has found yet more dirt to shift from one corner of the site to another and off-site there has been a big push to finalise the M&E (mechanical and electrical) drawings, i.e. the location of every pipe, wire, duct, valve, socket and light fitting in the house.

One aspect of Tree House I have long laboured over is the lighting. Of all the magical experiences offered by trees, the vision of sunlight through the canopy on a summer's day is one of the richest. Following this lead, I am keen to create an experience of light in Tree House that is truly out of the ordinary. For daylight, this involves large high-performance windows, contemporary stained glass, bright water and an abundance of foliage. For artificial light, the challenge has been to create an effective and exciting design that does not compromise our energy-efficient goals.

The central energy-saving message about lights has always been "turn them off". I won't argue with this, but as we have no Amish aspirations I prefer "turn them on but make sure they are doing the right job". Eco-lighting begins with design: thinking through exactly what you want from lighting and choosing the right bulbs, shades and fittings to achieve this. Lighting design is typically divided into ambient, task and accent lighting. Poor (or non-existent) lighting design often results in a lot of ambient lighting – bland, bright spaces – at the expense of good task or accent lighting. Ceilings studded with dozens of 40W halogens rarely create interesting or effective living environments and burn huge amounts of power.

Whatever your lighting design, make sure the technology is energy-efficient. The common incandescent bulb is the apotheosis of hidden domestic eco-horror as only 2% of the fuel energy burnt in the power station ends up as useful light from the bulb, the rest of the energy being thrown away as heat either in the power station or by the bulb itself. Unfortunately, although their inefficiency and short life make them costly, their cheap unit cost keeps them the nation's favourite. Nationally domestic lighting is responsible for around 6% of total electricity consumption.

Although light-emitting diodes (LEDs) are doing well and grabbing the headlines, compact fluorescents (CFLs) are still the most efficient bulbs available. Happily, you no longer have to settle for a bulb that looks like an egg whisk: modern CFLs come in all shapes and sizes including candle bulbs and miniaturised halogen replacements. One 7W CFL replaces a 30W halogen and will last seven times longer – a no-brainer if ever I saw one.

All CFLs run on mains voltage, so you can't do a straight bulb swap with low-voltage lights. You can, however, replace low-voltage lights with LEDs. These are

still fairly new to the market and do not always produce a high quality light but they are worth investigating as they use a lot less energy. Alternatively, if you want to stick with halogens, try the low powered (10W) capsules and avoid the greedy 50W lamps.

Alternatively, there's the Amish option: oil lamps. Burning the fuel *in situ* avoids all the energy losses of central electricity generation, but the palette for the lighting designer is rather limited. No good for us urbanites: although Harrison Ford looks good in a hair shirt, we composted ours a long time ago.

*Compact fluorescent uplighters cast a delicate
shadow pattern on the study ceiling.*

*The tree breaks into leaf in the bright spring sunshine.*

*design detail*

## Murano glass wall light

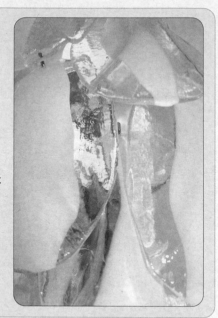

The best lamp shades do not shade the lamplight but transform it.

This 1960s Italian wall light is made from thick leaves of Murano glass, each with a frosted stripe down the middle. The three compact fluorescent candle bulbs in the fitting are disguised but their light is refracted and diffused beautifully through the glass leaves, creating a bright, glare-free and magical light-cast for the staircase of Tree House.

*resources*

The lighting design for Tree House was supported by Light Projects (020 7231 8282, www.lightprojects.co.uk).

The compact fluorescent lighting in Tree House, including 7W miniature ceiling-recessed downlighters, was supplied by Megaman (0845 408 4625, www.megamanuk.com).

Lighting is included in the Energy Saving Trust's database of energy-efficient products (0800 915 7722, www.est.org.uk).

### Design detail
Reclaimed and refurbished Italian wall light from 20th Century Collectables (01787 211450, www.20thCenturycollectables.co.uk).

### Publications
The following Best Practice in Housing publications are free to download or order from the Energy Saving Trust: *Cost Benefit of Lighting* (CE56), *Domestic Lighting Innovations* (CE80), *Energy efficient lighting* (CE61).

## 27 APR 05

We built as far as we could at the front of the house, leaving the back to catch up when the steel arrived. The exterior walls were made from the specialist engineered I-beams, but the few interior walls were knocked up with ordinary timber studs. The job was tricky because parts of the draught-inhibiting air barrier had to be installed at the same time as the structural components. Normally this wrapping goes on later when the structure is up, but a more careful integration of structure and wrap makes the building more airtight.

As an old-style Segal method man, Steve was not very convinced by the plastic and paper bags enveloping the building and took some convincing that they were worth the effort. Happily, when we eventually got round to pressure-testing the building, we found he had done a good job.

# Airtightness and ventilation

What's the difference between a hat and a scarf? I'm being strictly functional here, so ignore the finer points of fashion and consider the basic role these garments play in keeping us warm. A woolly hat provides a layer of insulation for your glowing scalp. Although heat is conducted through the hat, this takes time so your head warms up. A scarf does a similar job for your neck but this is secondary to its main function of plugging the air gap at the top of your coat. By making your clothing airtight, a scarf helps to reduce your ventilation heat losses.

*The Tyvek air-barrier is wrapped around the wall and floor junctions.*

> *The few interior walls are made from ordinary timber studs rather than the specialist I-beams of the main joists and walls.*

Houses conserve heat in exactly the same ways. Think loft insulation for woolly hat, cavity wall insulation for coat, and for the scarf – well, what? Draught-stripping is part of the answer but this will not make your home airtight as there are lots of tiny air pathways through your walls and roof through which heat can escape. It's possible to plug these holes too, but it's not easy. In the last week our little project team has been working out exactly how to install the two bags that will wrap up Tree House, one near the inside of the building envelope that will keep air and moisture out of the walls and one near the outside that stops air escaping but lets water vapour through so that any moisture that does get in the walls can be released (an important consideration for a timber-frame house).

Such airtight ambition has a potential downside. As your mother told you when she dandled you on her knee, never stick your head – let alone your house! – in a plastic bag. Sure enough, there is growing concern about the air quality inside some well-insulated modern homes. There are lots of toxic chemicals we bring in to our homes that will build up if ventilation is poor and any fire or boiler that is starved of oxygen will produce deadly carbon monoxide. Furthermore, whenever you make a cup of tea or have a shower, you are emitting a seemingly innocuous but dangerous gas: water vapour. A warm, humid house is a perfect home for dust mites, whose droppings are implicated in the huge rise in childhood asthma.

The answer to these problems is controlled ventilation that supplies enough air to keep you healthy with minimum heat loss. In our airtight house we will achieve this with whole-house mechanical ventilation, a box of tricks that extracts air through hidden ducts from the kitchen and bathroom and brings fresh air into the living room and bedrooms. Although they don't mix, these two streams of air cross each other through a fine mesh so that the heat from the exhaust air is transferred to the incoming fresh air. The fans in the unit require electricity to run but happily Vent-Axia's 'Lo Watt' system recovers a lot of heat energy for relatively little electrical energy input.

Whole house ventilation with heat recovery is tricky to retrofit but is definitely worth considering if you are building new, doing a major refurbishment or have problems with asthma. Whatever your circumstances and budget, reducing uncontrolled ventilation by draught-stripping, sealing holes where services penetrate walls and adding tightly packed insulation under suspended timber ground floors will pay you back handsomely in improved comfort and lower bills. You can then introduce extra fresh air only when and where it's really needed.

Most modern windows come with trickle vents: little grilles on the frame that can be opened just as far as you want. If you have these on your windows, make sure they are not giving you more fresh air than you really need. Extractor fans in kitchens and bathrooms are important for removing stale air quickly at the right time, but don't let them run when there is nothing important to extract. Above all, make sure rooms with open-flue fires are properly ventilated.

Unfortunately my current obsession with the thermal performance of Tree House has not been reflected in sartorial diligence: last weekend I found myself on a windswept Scottish hillside without hat, scarf or even a decent coat. My well-dressed companion, two-year-old Talia, was considerate enough not to scoff from beneath her strawberry-red woollen hat. Give her a year and I have no doubt she'll be sitting on dizzy uncle Will's knee telling him all about the dangers of plastic bags.

*The heat-recovery ventilation unit was intriguing, but had to be put back in its box until the house was airtight.*

## Aerating shower head

A little extra oxygen can make a significant difference to your water consumption. Taps and shower heads with aerators mix water with air at the outlet, creating a soft but full flow for much less water throughput. This aerating shower

provides a gentle, refreshing alternative to the extravagant inefficiency of a power shower.

*resources*

The whole-house heat-recovery ventilation system for Tree House was supplied by Vent-Axia (01293 530 202, www.vent-axia.com).

### Design detail
Raindance Puro-Air shower by Hansgrohe (0870 7701972, www.hansgrohe.co.uk).

### Publications
The following Best Practice in Housing publications are free to download or order from the Energy Saving Trust (0800 915 7722, www.est.org.uk):

*Energy efficient ventilation in housing* (GPG268)

*Improving air-tightness in dwellings* (CE137/GPG224)

# 04 MAY 05

**Immersed in the day-to-day details of building Tree House, it was easy to forget just how much thought and effort had gone into designing it. In this diary entry I took a step back and reflected on the many ways in which nature and ecology influence architecture and design. In doing so, I wanted to stress that eco-building is something much bigger than a collection of technical issues.**

**This was also an opportunity to acknowledge the all-important creative role played by our architect.**

## Architects and eco-buildings

This morning I turned on my computer and downloaded seven overnight emails from our hard-working architect, Peter Smithdale. Among other things, they concerned ventilation ducts, underfloor heating controls, coping-stones for our garden walls and shower drainage. In principle, our contractor is building a house that Peter drew seven months ago but in practice there is an endless stream of details to resolve, a process that potentially includes negotiations with contractor, site foreman, engineer, suppliers and (not least) an ambitious client with a time-consuming habit of pushing everything beyond standard practice.

*Architect Peter Smithdale considers his drawings with Steve.*

We feel very lucky to have had Peter's professional support over the last two years as he has achieved that rare balance of listening and challenging, empathy and imagination. Every one of our ideas has been treated with critical respect so we feel deeply involved in the design of the house. Although Peter is a member of the Royal Institute of British Architects, he does not conform to the stereotype of the profession expressed by the alternative meaning of RIBA: 'Remember – I'm the bloody architect!'

It helps of course that Peter shares our interests. We approached his firm, Constructive Individuals, because they were ecological architects with experience of working with self-builders. This was a good place to start, but what we've really valued is Peter's willingness to look beyond the technical aspects of our zero-carbon brief and work with our metaphor, to help us design a house that captures the spirit of a tree.

Although the question of what 'counts' as ecological building tends to focus on the details of construction and performance, there are many wider issues raised by this concept, not all of which are in the power of the architect to resolve. For example, is a house 'green' if the people who live in it still squander resources and burn fossil fuels in every way they can? Here we certainly hope to follow the lead of BedZED in south London where very high performance housing is integrated with live-work spaces, car clubs, local food schemes, recycling and public transport.

At another level, ecological building can be understood as a creative response to the natural world. This, very much the architect's domain, may mean integrating buildings with their landscape and setting, exploiting organic forms in the shape of the building, or drawing on deeper principles of nature's design in the philosophy of a project. With Peter's support, we have sought to capture all these qualities in Tree House. The house is a respectful response to the mature tree that dominates the plot; the form of the tree is subtly picked up throughout the macro and micro design; and, above all, the house will work like a tree, drawing all its energy from the sun and thriving in harmony with its natural environment.

We were encouraged to imagine a house that captured the form and beauty of a tree by a building that Peter introduced us to: Alvar Aalto's Villa Mairea. A year ago we embarked on an odyssey by sea and road to a remote corner of Finland to visit this remarkable house, which combines the discipline of modernism (it was built in 1938) with a richly detailed response to the forest in which it is built. It may not be an eco-house by today's technical standards, but its integration in its natural setting and subtle evocation of the forest in its design provide a fine example of what a modern response to nature can achieve.

Frank Lloyd Wright, whose Falling Water is another profound response to a natural setting, once said that physicians could bury their mistakes whereas architects could only recommend the planting of vines to their clients. I have already bought the vines – *Ampelopsis megalophylla* to be precise – but Peter can rest assured that these vines are an integral part of our summer shading strategy and most certainly not a veil of embarrassment.

*The first joists are installed at the front of the house.*

*design detail* Pergola

One of the more explicit ways in which Aalto echoed the surrounding forest in the design of the Villa Mairea was his use of roughly finished poles for various external details including this pergola.

Although our design focus is more on a single tree than on a forest, we are incorporating a pergola to a similar design along and above the pond that runs the width of Tree House, dividing the living space from the garden. This pergola, made from coppiced sweet chestnut, will shade the room in summer and reduce the risk of overheating, as well as helping to blur the inside-outside distinction of our living space.

## Constructive Individuals

Constructive Individuals is a small firm with a portfolio of interesting ecological architecture largely undertaken with communities and self-builders.

Tree House began its design life in the enthusiastic care of Simon Clark, whose organic, curving stair tower became a central feature of the romantic design. When Simon left the firm to travel the world, Peter Smithdale took over. He nurtured our ideas with a generosity and critical intelligence that proved invaluable.

Peter's colleague and professional sounding board is John Rees, whose work includes Ben Law's Woodland House, a very low-impact house built in a Sussex wood for next to nothing.

The Royal Institute of British Architects has a database of members and practices (020 7307 3700, www.architecture.com).

Tree House architects: Constructive Individuals (020 7515 9299, www.constructiveindividuals.com).

To find an architect with green interests, begin with the membership of the Association of Environment Conscious Building (www.aecb.net). Member information is available through the online database or in the Green Building Bible (K Hall, Green Building Press 2005, www.newbuilder.co.uk).

### Publications
*Alvar Aalto: Villa Mairea* (Alvar Aalto Foundation / Mairea Foundation 1998). A rich account of an exceptional house.

*New Organic Architecture: The Breaking Wave* (D Pearson, University of California Press 2004). An exploration of buildings that are inspired by the forms and flows of nature.

# 11 MAY 05

The experience of self-build is dominated by the daily struggle with the structural guts of the building and the services that run through them. Yet the experience of living in a house is profoundly affected by the finishes that cover up all the mess. Although these finishes are usually installed in the dying hours of the build, it's crucial not to leave their specification too late.

One of the pleasures of building Tree House was the search for beautiful, ecological floor and wall finishes. Of all the journeys I made in this quest, the one described in this diary entry was by far the most rewarding and enjoyable.

*The restored fell-side on the approach to the Kirkstone quarry.*

# Stone

I'm standing in the rain on a steep fell in Cumbria, watching the clouds curl round the cliffs high above me. A scree spills down the hillside: a cascade of boulders bursting through the soft green earth. Sky, rock and land are united in the distinctive grey-green tones of the Lake District. The scene ought to be elemental, but the fleeting sight of a large earth-mover high up on the fell-side suggests that all is not what it seems.

I am in fact climbing the approach to the quarry on the Kirkstone Pass, a guest of the manager, Nick Fecitt. I'm here to check out the environmental credentials of one of the key materials in the palette of Tree House: Kirkstone slate. We are off to a good start, for the hillside is actually a restoration, an old quarry scar filled with stone waste then carefully seeded to recreate the appearance of the surrounding fells. Given the location of the quarry in the middle of a national park, this work is essential to sustaining a long-established industry in an area where tourism now has first priority.

Kirkstone slate is principally used for high quality floor and wall tiles and for work surfaces. The 'Silver Green' slate is the most distinctive in the range as every tile is patterned with fine bands of grey and green, seemingly cut against the grain of the rock. At the top of the quarry, where huge boulders are being dragged out of

*The stone is pulled from the quarry face.*

the mountainside and split by the earth-movers, Nick explains their origin. The stone was formed by a series of volcanic eruptions, each of which created a new layer of mineral and ash deposits. Then, in the collision of the islands that preceded Scotland and England, this rock was thrust upwards and compressed against the grain of the original sedimentation. Consequently, the rock now cleaves at right angles to the volcanic layers, exposing the early geological history of the rock on every tile.

The timeless beauty and strength of slate are its core eco-credentials. Like wood, most stone has an appeal that suffers little from the vagaries of fashion and so typically enjoys a long life once installed. This is only possible because it is so hard-wearing and durable: even if your floor does get ripped up by future inhabitants with dubious taste, the stone will hopefully achieve a long life via the reclamation yards.

Nick shows me the progression of the slate from cliff-face to consumer product, a journey through a series of bleak sheds where huge circular saws cut and hone the rock, first into blocks and then into tiles or larger slabs for work surfaces. Although the transformation is remarkable, I am struck by how simple the process is, requiring relatively little energy and no toxic extras. This is a major environmental advantage of stone: the tough part of the manufacturing process has already been done by Mother Earth. Unlike concrete, bricks, ceramic tiles and synthetic finishes, no further firing is needed beyond that completed several million years ago.

As we drive back to the company office in the valley below the pass, Nick draws my

*Nick Fecitt draws attention to the extraordinary geological detail exposed in the slate.*

attention to the local buildings. Although slate still defines the character of the Lake District, its ubiquitous use in earlier buildings is not matched in many more recent constructions where block walls and concrete roof tiles are also in evidence. The UK building industry has long departed its local roots, to the extent that using British (let alone local) stone is exceptional. Unfortunately the eco-profile of stone drops dramatically as soon as you start importing it, given the energy required to transport it. South London may not be Windermere but a truck on the M6 is a relatively small eco-cost to pay compared with ships from the South Seas.

Kirkstone isn't cheap but it is gorgeous: I am confident that its unique character will pay long-term dividends. If you're in the market for a high-quality durable finish, think twice before you shell out for Indian sandstone or Italian travertine and consider buying a slice of an English national park instead.

*The living room floor looks at its best in the dappled afternoon light.*

design detail

## Kitchen surface

Our slate ground floor is complemented by a kitchen worktop that also began life in the Kirkstone quarry. This is Kirkstone Sea Green slate, which has a more consistent colour than the livelier Silver Green that we used on the floor.

We originally specified wood for our kitchen worktop but, given the amount of oil and water we like to splash around our kitchen, we decided that the maintenance demands of a timber surface would be too great. Slate is robust, beautiful and maintenance-free.

resources

For a Kirkstone catalogue, phone 01539 433296 or see www.kirkstone.com.

### Publication

*The Green Building Handbook* (T Wooley, S Kimmins, P Harrison and R Harrison, E & F N Spon 1997). This comprehensive guide to the environmental impacts of construction materials describes the benefits of stone as a low-impact heavy-weight material.

# *18 MAY 05*

For several weeks I held on to the hope that the critical steel components would arrive on site and I could continue my story seamlessly. Regrettably this was not to be, so in this diary entry I decided to come clean.

Undaunted by the slow progress on site, I continued my tales of sourcing eco-products. If my trip to the Kirkstone Pass was the most enjoyable of these escapades, the experience recounted here was the most infuriating. I remain shocked that despite all the political talk of tackling climate change and promoting energy efficiency, information about the power consumption of electronic goods remains remarkably hard to come by. What's more, nobody seems to care.

Back on site the building continued to rise, albeit only at the front of the construction where no bits of steel were needed to keep the walls up.

## Television and electronic goods

I'm looking out of our bathroom window for the first time and loving it: the rising multiple trunks and lush foliage of our tree complemented by a clear view of one of Clapham's more unusual buildings, an elegant Wesleyan mission hall with a striking triangularity of form. Unfortunately the bathroom window itself is missing, as is the ceiling, the floor above and the roof above that.

*Every vertical wall stud is carefully positioned.*

Once again we're way behind schedule thanks to some unruly steel that has twice been sent back for refabrication. The house appears to mount a vigorous immune response whenever this thoroughly un-eco, un-tree-like component is thrust into the heart of our timber frame.

*As the tree fills out, there is still little sign of the house.*

I could weep into my fairtrade coffee but instead I am focussing on some of the finer details of our 'zero carbon' specification. The energy consumption of everything in the house has got to be kept as low as possible in order that our total demand does not exceed what we generate on our roof. So this week I headed for London's west end in search of another domestic view, the all-important window on the world offered by our beloved television sets. Could the finest of London's retailers tell me about their energy consumption, I wondered?

First stop, John Lewis, where I was assured that all televisions had 'minimal' electricity consumption so I needn't worry. The same adjective was used by the impatient assistant in House of Fraser, who compared televisions to irons, evidently a symbol for him of domestic inconsequence. In fact, we would need a clutch of extra power stations if televisions performed as badly as power-hungry irons.

The cheerless assistant in Selfridges began in the same vein but did tell me that plasma televisions use more electricity than either traditional cathode ray tube (CRT) sets or liquid crystal display (LCD) flat screens. He then showed me three Sharp LCD televisions that actually had energy labels stuck to them, albeit so discreetly that they needed pointing out. He evidently felt that this information would adequately represent the rest of his stock.

In Tottenham Court Road, I tried a Sony specialist where my questions were met with incredulity. Power consumption? Was I going abroad? Finally, a few doors up at PNR Audiovision, patient and cheerful Minesh finally took my questions seriously

*The view from the first-floor bathroom window.*

and did a good job of answering them, even downloading and printing out information from the internet to enable cross-brand comparison. He showed me the labels on the Sharp televisions but didn't assume they were representative of all brands.

For the record, televisions do not consume 'minimal' electricity. Screens have grown and stretched, as has our time in front of them (and perhaps our bellies too), so they take a fair slice of the domestic energy cake. The bigger your screen, the more power you draw, but power consumption also varies across technologies and brands. As plasma screens are shameless energy-guzzlers, the competition is really between LCDs and traditional CRTs.

Although LCDs have gained a reputation for energy efficiency, this is only a sure bet for smaller, computer-sized screens. Above 18 inches CRTs are still winning, though not consistently. For example, a 32inch Sharp widescreen CRT television has a power consumption of only 85W, compared with 135W for a flat LCD screen of the same size. But this isn't really comparing like with like: a 'pure flat' Sharp CRT has an even higher consumption of 160W (what a difference a little curvature makes). Given that the future is flat, it looks like LCDs may be the long-term favourites after all.

Hopefully Sharp's energy labelling will be followed by other brands, so that quick comparisons can be made on the shop floor. The leading brands are committed to this so it's time they delivered. Remember to check the standby consumption too, which ought not to be more than 1W, as even the most diligent among us still forget to get up and turn the thing off.

If you're interested in the energy label on a fridge, take the same interest in any electronic goods you buy. If a few more people start asking these questions, those high-street looks of patronising incredulity might slowly begin to fade.

*Ford looks in vain for an energy label in a sea of electronic goods.*

*design detail*

## Laptop computer

Laptop computers are marvels of energy efficiency, using far less power than desktop models. As battery life is an important part of the specification of any portable computer, there has always been an incentive to keep laptop power consumption low. Perhaps we should all run our houses and cars on batteries: the same design incentive would soon bring the energy-efficiency step-change the world needs.

## resources

To find out the power consumption of televisions and other electronic goods, check out the company websites, look at each individual product and check the fine detail of the specification. Standby power consumption on all electronic goods ought to be no more than one watt.

Britain is currently struggling to implement the EC Directives on Waste Electrical and Electronic Equipment (WEEE) and Restriction of the Use of Certain Hazardous Substances in Electrical and Electronic Equipment (RoHS). As WEEE requires producers to take responsibility for recycling their products when consumers have finished with them and RoHS limits the materials that can be used in their manufacture, the impact of the directives on the toxic stream created by the huge demand for electronic goods is likely to be considerable. Information about these directives is available from the Department of Trade and Industry (www.dti.gov.uk/sustainability).

# 25 MAY 05

The journey recounted in this diary entry was the perfect antidote to my frustration with our slow progress on site. The bright freshness of May, the spring sunshine and the enfolding landscape of the Sussex Downs provided an idyllic setting for a series of encounters with wood, the principal construction material of Tree House. I returned to Clapham with my long-term vision of a beautiful and characterful building fully restored.

# The right wood for the job

Every piece of wood tells a story and, in this column, every story comes with a ha'penny-worth of eco-enthusiasm. So here are three woods, three stories and one and a half pence-worth of my weekly eco-evangelism.

Last week I spent a bright and beautiful day sourcing various timber details for Tree House in the South Downs. I was accompanied by our architect, Peter

Peter gets to grips with the salvaged oak joists.

The joists will frame the windows as the branches of a tree frame the sky.

Smithdale, and we were both exhilarated by the prospect of some intimate encounters with our mutually favourite building material. There's nothing like meeting the trees your wood comes from, whether or not you are inclined to hug them.

Our first stop was not, however, a forest but an early eighteenth-century oak barn mid-way through conversion to a chunky des res. Many of the timbers were being reused but there were a good few left over. Andy the builder led us through the farmyard to a dark shed where a large pile of old oak joists lurked in a corner. We had found what we wanted: beautiful, characterful timbers that will frame our windows, like branches framing the sky through the canopy of a tree.

Although their function will be purely aesthetic – our actual window frames will come with our high performance windows – they will symbolise the continuity of timber buildings and the ecological value of reusing their component parts. Although we are inspired by the nouveau-riche sycamore that dominates our plot, it will be a pleasure to incorporate a little aristocratic oak into the fabric of the building.

Having made our choices from amid the pile, we headed west to one of the biggest broad-leaf forests in England where woodsman Robin Carter of FSC-accredited Timber Resources International became our guide. We were looking for two tree trunks which, debarked, will rise up inside the staircase of Tree House,

*To sustain the biodiversity of the woodland, only small areas are felled at one time.*

providing an immediate, tactile evocation of the form and character of trees right in the heart of the house.

Before long we were driving deep into a forest of super-straight Douglas fir in search of fallen trees, as a tree that has been dead for some time has a head start in drying out over a newly felled tree. We want our tree trunks sooner rather than later, but we also want to avoid the nasty chemical baths that can short-cut the curing process, so the drier they are to start with the better. Nonetheless, the poles will split and crack over the years as they slowly contract, a character detail that we have decided to embrace.

The principle here is to learn the strengths of wood and work with them, rather than assuming that you can always get a material to do what you want by using a toxic enhancement. This principle casts a shadow over the ubiquitous use of timber preservatives, which are regularly slapped on woods of all kinds without much thought. The most ecological approach to timber preservation is always to avoid chemical treatments altogether by building carefully, paying special attention to how wood is ventilated, and using the right wood for the job. There are plenty of wooden houses throughout the world that have survived centuries without a drop of preservative.

The final item on our arboreal shopping list was poles for a pergola, designed to reach across the pond that will run the width of our main living space. For this, Robin took us to another corner of his thousand-acre wood, to a plantation of coppiced sweet chestnut. Sweet chestnut is a highly durable wood (which certainly doesn't require preservatives) and coppicing is a particularly sustainable way of managing woodland because you never actually kill the tree.

*Peter and Robin inspect a Douglas fir tree trunk, destined to hold up the staircase of Tree House.*

Oak, Douglas fir and sweet chestnut. Recycle buildings, treat wood with care rather than chemicals and keep the forests alive with sustainable management. Good value, I reckon, from one brief but glorious day in the rolling Sussex countryside.

## design detail

### Woodpile fence

This boundary is an integral part of our wildlife garden, turning the valuable habitat of a woodpile into an architectural feature. The logs are from a poplar tree that grew in the next-door garden to our Brixton flat that had to be felled because of subsidence problems. I managed to divert the logs from the chipping machine just in time. In due course, we hope this boundary fence will be overrun with a rich diversity of flora and fauna.

## resources

The staircase newel posts for Tree House were supplied by Timber Resources International (01428 741349, www.timberresources.co.uk).

The Forest Stewardship Council in the UK provides information on sources of sustainable timber and timber products (01686 413916, www.fsc-uk.org).

To source reclaimed wood, see the directory of salvage yards maintained by Salvo (www.salvoweb.com).

If you are interested in learning more about the protection and management of British woodlands, contact the Woodland Trust (01476 581135, www.woodland-trust.org.uk) or the Small Woods Association (01743 792644, www.smallwoods.org.uk).

# *08 JUN 05*

In June, the house finally took off. The steel arrived and fitted, the sun came out and stayed out, and a flurry of carpenters cultivated impressive tans as the joists and wall studs were meticulously prepared and installed.

Unfortunately the long delay did mean that our windows arrived well before the frame was ready to fit them. This was a little worrying given the risks of storing anything fragile on a small and busy building site. As these were no ordinary windows – made-to-measure, triple-glazed and shipped from Sweden – they seemed to cry out for falling hammers. Long after the windows had been installed, one such hammer did finally hit its target.

## Windows

Our windows arrived this week: a consignment of high performance glass and timber shipped all the way from Scandinavia by Swedish Timber Products. As they will be integral to both the environmental performance and the human experience of Tree House, we cut no corners here. Windows are magical: they breach the boundary between inside and outside without sacrificing the security and shelter that define our experience of home. And if you're investing in magic, there's really no room for compromise.

*The steel is finally put in place.*

Windows play a complex eco-role. They bring in the warmth and light of the sun and so can reduce demand for heat and power but they are also remarkably good at chucking heat away. Lots of glazing can also lead to over-heating and reliance on energy-guzzling air-conditioning, an increasing risk in our warming climate. The optimal eco-window is therefore a matter of good building design as well as good window technology.

Although both of the main living spaces in Tree House have substantial glazing, each room is protected from overheating by careful shading: external Venetian blinds on the ground floor and the tree itself on the top floor. The tree contributes its own magic to the design, shading the house in summer and letting light through in winter when extra warmth is most needed.

We can only incorporate these outside-in spaces into our ultra-efficient house because the windows themselves let very little heat escape, in fact less than a fifth of the heat lost by an ordinary sash window. This is achieved by tackling all three modes of heat flow: conduction, convection and radiation.

Conduction is the direct flow of heat through a material, such as through the handle of a poorly designed pan on a stove. This is how heat moves through a pane of glass, so if you add more panes – our windows have three – the heat loss takes longer.

Convection takes place when heat is carried along by the movement of a liquid or gas. The swirling of the water around the potatoes in my pan is evidence of such convection currents. The heat transfer from pane to pane in double glazing is largely due to convection currents in the air gap, so with a heavier gas – our windows are filled with argon – the currents slow down.

*Steve's smile indicates that all is well with the measurements this time.*

*But Steve is less happy when the windows arrive before they can be installed.*

Radiation is a universal phenomenon: everything radiates heat. Even when I have turned the stove off and served my potatoes, I can tell the ring is still hot because of the heat it radiates. Every window pane radiates heat out to the sky but different materials do this at different rates: our windows have two invisible 'low-emissivity' coatings that reduce this heat loss.

Finally, because our windows are extremely airtight, they will lose a minimum of heat from draughty joinery, a major and uncomfortable problem in older houses.

Window frames are also an important eco-issue. It's something of a disaster that u-PVC windows have triumphed in the market, given how ugly they are and how much toxic waste is created in their manufacture. Why use such a poisonous material when you can use wood instead? Happily, Swedish Timber Products has FSC accreditation, so we can even be sure of the sustainable management of the forests from where our windows began their journey to Clapham. The only eco-downside is the boat trip from Sweden, but this is a small price to pay for the efficiency savings that the windows will bring over the lifetime of the house.

By the way, my cooking analogy is not entirely accurate. For a start, our ultra-efficient AEG-Electrolux induction hob generates heat in the pan but doesn't get hot itself. Furthermore, I'm on a low carbohydrate diet and have long since given up cooking potatoes. But that is definitely another story.

*The magic of windows – Ford relaxes with the cats in the inside-outside living room.*

*design detail*

## Reclaimed stained glass

The constraints of the site meant that Tree House could only have one south-facing window. The constraints of the planning approval required this window to be opaque, as it overlooks a neighbour's garden. It is therefore made from very thick glass blocks, producing a bright but diffuse light.

As there is no view from this window, we have used two salvaged stained glass panels to make the light more exciting. I found these panels in a rather upmarket reclamation yard in north London that specialises in making creative use of salvaged materials. I noticed straight away a certain tree-like quality to the design and bought them on the spot.

*resources*

The windows for Tree House were supplied by Swedish Timber Products (01347 825610, www.swedishtimberproducts.co.uk).

The roof windows were supplied by the VELUX company (0870 240 0617, www.velux.co.uk).

### Design detail
Reclaimed stained glass from Retrouvius (020 8960 6060, www.retrouvius.com).

### Publications
*Solar Energy and Housing Design* (S Yannas, Architectural Association 1994). A detailed guide to the many ways houses can benefit from the light and heat of the sun.

The following Best Practice in Housing publications are free to download or order from the Energy Saving Trust: (0845 1207799, www.est.org.uk):

*Benefits of Best Practice: Windows* (CE14)

*Windows for New and Existing Housing* (CE66)

# 15 JUN 05

By mid-June the outline of the house had emerged and it became possible for the first time to get a sense of the rooms-to-be in real space. Unlike masonry houses, timber-frame buildings remain transparent when the primary wall structure is erected, so we could still see all the way up and through the house even when the top floor was being built.

From this point in, there was a new level of excitement about the build: every step took us closer to the interior spaces that we had imagined for so long. The daily frustrations on site, such as trying and failing to get utilities connected, now felt more like a burden worth bearing.

*The skeleton of the house rises towards the tree's canopy.*

# Green electricity

How many builders does it take to fit a lightbulb? If you don't have a power connection to start with, there's one to dig the trench, one to dig the hole in the pavement, one to connect the cables, one to fill the hole, one to fit the meter, one to connect the meter to the consumer unit, one to certify the connection, one to install the lighting circuit and one to throw a complete wobbly when the bulb fails to light.

Each of these events has to be carefully scheduled: if you miss one you're in trouble. Although our power supply should have been operational this week, an unexpected school run meant that foreman Steve missed the appointment with the connecter and everything else unravelled.

Without this connection, Steve is stuck with our noisy on-site generator and regular trips to the petrol station with his jerry-can. Not that power from the Grid is likely to be any more eco-friendly than diesel: London's electricity is now principally supplied by the French company EDF, so the next stage of the construction of Tree House may well be nuclear-powered – low carbon, perhaps, but as tricksy for future generations as any fossil fuel.

Happily within a few weeks the photovoltaic panels on our south-facing roof will be in place and our little urban power station can be switched on, providing up to 5kW of solar power for the last stage of the build. In practice, however, a lot of the power generated on our roof will be surplus to Steve's requirements and duly exported to the National Grid (as it will be on summer afternoons even when the house is occupied). The power from little urban generators like us never gets anywhere near a high voltage transmission line and is simply soaked up locally, so whatever deals our neighbours may have with their electricity suppliers, their washing

*Chippies are suddenly in great demand on site. Tom joins two I-beams for an extra strong wall stud.*

machines and daytime telly will actually be driven by our eco-friendly electrons.

The business of electricity generation, purchasing and consumption is a bit mind-boggling, not least because electricity is so inscrutable. The key point to bear in mind is that you don't pay your electricity supplier for the actual electrons you use; you pay them to put power into the Grid to match your consumption. Your supplier cuts deals with the electricity generators within a complicated trading system that is designed to meet national demand through forward purchasing. So if you have a 100% green electricity supplier, such as Ecotricity or Good Energy, you are paying them to buy power from renewable generators but what you actually turn into heat, light and Big Brother remains cloaked in subatomic secrecy (unless, like us, you're making the stuff yourself).

Whatever you do, if you switch to a green electricity tariff don't assume that you no longer have to worry about how much power you consume. If you start leaving the lights on, this extra load will almost certainly be met by burning extra coal, gas or uranium, especially at times of peak demand. Currently the renewable energy generation capacity in Britain is very small, so although it's good to buy into it and support its expansion, squandering green energy is just as bad as wasting old-school dirty power.

*Steve prepares the service ducts for the electricity and water companies.*

A future-friendly power infrastructure is only achievable if we complement renewable energy with radical energy efficiency. Tree House is an expression of this principle writ small. I knew at the outset that we would never be able to generate enough heat and power on our little London plot to meet the typical energy demand of a 2.8 person family (two men, four sedentary cats). Our goal of generating more energy over the year than we consume is only achievable because our rooftop power station will sit above an exceptionally energy-efficient house. I hope that makes some sense. I'm aware that electricity can seem about as intelligible as reality TV. A mysterious energy source for an even more mysterious world.

*The form of the stair tower, the trunk of Tree House, emerges.*

## Herb garden

Although the power we generate on our roof is exported to the Grid when we cannot use it directly, we do have means of storing solar energy on-site for later use. The simplest of all is our herb garden, a collection of square pots in a sunny corner at the back of the house where sunlight is transformed into exquisite carbohydrates.

Anyone can grow herbs – all you need is a window ledge. As air-freighted supermarket herbs have huge hidden energy costs, this is not a trivial way of saving energy. Many herbs are also excellent sources of nectar for bees.

The solar photovoltaic panels for Tree House were designed and installed by Solar Century (020 7803 0100, www.solarcentury.co.uk).

Almost all electricity suppliers offer green tariffs but what this means in practice varies. Some buy all their energy from renewable generators, others are investing in new renewable capacity. Green Electricity Marketplace (www.greenelectricity.org) and Energylinx (www.energylinx.co.uk) offer area-based comparisons of the tariffs of renewable electricity suppliers.

The Energy Saving Trust offers advice about combining domestic renewable electricity generation with appropriate energy-efficiency measures (0800 915 7722, www.est.org.uk).

## 22 JUN 05

The timber frame quickly smothered the steel and turned the long months of groundworks into a fading memory. It was a huge pleasure to go to the site every day and witness the wooden outline of the house emerging next to our tree. The erection of the frame was not exactly rapid as every joist and wall stud had to be cut to size on site, but after so many delays the relative speed was exhilarating.

This felt like a good time to acknowledge the support and consideration of our neighbours who had lived with a building site for nine months and could plainly see that we were still far from completion. Paula, Mark, Thomas and Gabriel; Pam and Doug; Joss, Larry, Lucas and Jasper: thank you!

# Good neighbours

It's nine months since Steve and George first set foot on our patch of overgrown land in Clapham and began cutting back the undergrowth beneath our big old sycamore. If everything had gone to plan, Ford and I would now be settling into our beautifully formed eco-environment and watching our four cats negotiate their airtight cat doors and the pond beyond (feline stepping-stones will be provided).

Instead, we don't even have a roof above us and our hopes of moving in before the end of the summer are fading fast. Despite the huge number of things still to do, we might yet get in before October because once the house is sealed lots of different trades can get to work at once, but this assumes a blessing of good luck and sharp organisation that we have not so far been graced with.

Mark, Thomas, Gabriel and Paula Wilson, our neighbours from heaven.

*The first-floor joists are hung off wood packed into the steel beam.*

*George looks after a special visitor to the scaffolding – a south London stag beetle.*

We may be behind schedule but we're still building a beautiful ecohouse on a difficult site in the centre of London, so I suspect we are unlikely to find many sympathetic ears for stories of misfortune. Furthermore, each time we have faced a serious delay – due to extortionate insurance, elusive timber and bodged steelwork – we have made the most of the extra time and kept our sights firmly on the vision of what will be.

Building a house is like a game of Scrabble: whatever letters you are dealt, you can still come out on top. The secret is to see the potential in even the trickiest of letter combinations and not to lose heart when the space for your seven-letter word is ruined by some inconsequential five-pointer. Although we've had our fair share of duff hands, I have kept my cool in the knowledge that a 400-point score is still possible as long as we don't start swearing at our vowels.

Taking the long view, our luck has been in more often than it has been out. Finding a building plot in Clapham within two days of starting the search was like drawing ZYMOTIC from the bag on my very first go. Inebriated by this unexpected turn of events, I was convinced that my next hand – talking to the neighbours –

*The outline of the first floor takes shape.*

would reveal a fist full of 'I's. Approaching each doorway with trepidation, I was prepared to encounter Clapham's most hard-bitten, intransigent inhabitants, then turned over my letters and found QUAICHS staring back – and toasted our good luck once again.

We have great neighbours. They have been helpful well beyond the call of good manners, not only tolerating scaffolding, generators and Steve's jokes, but also helping out with storage space, a water supply and ever-welcome encouragement. To the left, Paula and Mark Wilson have borne the brunt of the digging and skipping without complaint, though I fear their son, Thomas, may object when Tree House turns out to be rather different from the treehouse he is expecting. To the right, Pam and Doug Woodhead have considerately let us turn half their garden into a storage yard, a kindness that has made immeasurable difference to the daily business of getting round the site. Finally, Joss and Lawrence Walford at the back have simply been as enthusiastic and supportive as anyone could hope for, even when the diggers left and their young sons lost interest in the strange world beyond the garden fence.

For our part, we have sought to take their interests seriously at every stage of the game. Above all, this meant addressing their concerns at the design stage, an aspiration that proved to be far less onerous than expected once we began to see how their requests could work for us as well as for them. Perhaps this is the heart of neighbourliness: finding mutual gain through mutual respect, a principle that we hope to sustain for the next forty years or so. As any tournament Scrabble player will tell you, if all contestants aim to open up the board rather than blocking each other's chances, everyone gets to play their seven-letter words.

The frame is finally ready for the windows.

## design detail | Hidden gate

Our crazy back garden fence has a secret doorway into our neighbour's back garden. Even the handle merges with the torn and stressed metals of the fence.

As a gate had always been on this boundary, we didn't want to shut our neighbours, Joss and Larry, out of our lives when the new (but salvaged) fence went up, not least because they had been so considerate to us and supportive of our plans. Their two sons are still too small to operate the secret door. What games will they play when they finally work it out?

# 29 Jun 05

By the end of June the walls of the timber frame had reached their zenith: the peak of our big southern-pitched roof. The roof itself was still to come, as were the floorboards and the sheathing for the walls. As the building was still a mere skeleton, we could look down from the top of the scaffolding and see the form of all the rooms of the house below, a line drawing waiting to be coloured in.

Lots of glorious sunshine kept the mood bright among the assorted chippies on site. Some people came and went, others stayed. At this time, the regular workforce started to expand, with Steve and George joined by Pete, Genc and Nikolin, whose considerable skills would see the project to its close.

# Daylight

Do you use renewable energy at home? Whoever you are, wherever you are and whatever you think of wind turbines and solar panels, the answer to this question will always be 'yes'.

The renewable energy I have in mind is daylight, an extraordinarily important natural resource that profoundly shapes the design of our homes. It may not yet be customary to design houses that store solar energy for warmth or that convert solar energy into hot water or electricity, but no architect would dream of spurning natural light.

Daylight has qualities that can never be matched by artificial light. We value its vibrancy, its daily ebb and flow, its intimate relationship with the seasons, and its unpredictability. When daylight is plentiful, we are literally delighted; when it is in short supply, we get S.A.D. Yet despite being showered with these riches, we still don't make the most of daylight in modern house design, not least because easy-on electric lights have lowered the priority of daylight among the many competing concerns of architects.

Currently, Tree House has exceptional daylight as our unsheathed timber frame provides a line of sight right through the building and out the other side. When the walls get their plywood skin next week we will get a first indication of the

effectiveness of our ultra-
efficient Swedish windows in
lighting the interior.

But there is more to daylight
than decent windows. On the
ground floor of Tree House
our large open-plan living
space enjoys extensive
glazing, but this all faces
one way – west. Hopefully,
this daylight will still get
deep into the house, thanks
to the shortage of interior
walls, the use of light colours
for reflective surfaces and
the sparkling addition of the
pond that runs the width of
the house just beyond the
windows. When necessary,
our external Venetian blinds
will minimise glare without
throwing the room into
darkness.

*With no sheathing on the walls,
the building is still flooded with
natural light.*

Even with these measures, we
were worried that the kitchen
space at the back of the room
might be too dark to make
morning coffee without turning
the lights on. The solution has been to puncture the wall between the kitchen and
the small laundry/shower room beyond and so bring daylight into the room via the
east-facing laundry window. We are plugging the hole in the wall with something
rather special: a five-foot panel of contemporary stained glass, designed to
express the beauty of sunlight seen through the canopy of a tree on a summer's
day. Made by Sarah McNicol of Juicy Glass, this stunning panel has inevitably
morphed from a little eco-idea into the design centrepiece of our kitchen.

If you have dark spaces in your home in the middle of the day, avoid turning the
lights on by keeping surfaces light and bright, choosing window coverings that
can be adjusted to balance useful light with glare control and positioning key task
areas near to windows. The most important time to consider daylight is when you

are investing in major changes to your interior space. Then you can improve your windows, add roof windows (VELUX produce high quality, energy-efficient options) or even install daylight pipes that bring light deep into your house from your roof. This is also the time to fully rethink your household activities to better suit the sunlight (see the paintings of Vermeer for inspiration).

While you're about it, remember that daylight is as playful as it is eco-friendly. You may not want to live in Chartres cathedral, but stained glass, water and characterful shading can all play their part in making your living space simply divine.

*The architect's three dimensional daylight model of the top floor of Tree House shows the morning light flooding in.*

*The reality exceeds expectations.*

*Sarah McNicol's stained glass panel in the kitchen.*

## profile

### Sarah McNicol

Although our commission for the kitchen of Tree House was relatively small, Sarah brought all her artistic powers to bear upon it and came down from Nottingham, where Juicy Glass is based, to discuss the project and get acquainted with our tree.

Sarah's work is characterised by distinctive colours and the interplay of fragments of glass. She does not typically use recycled glass but nor does she throw anything away. Every splinter and shard of left-over glass is kept for the next project.

## Suncatchers

These stained glass suncatchers rise up out of our pond to catch the light from both above and below, throwing shimmering colours across our inside-outside living room throughout the year.

The contemporary stained glass panel for the kitchen of Tree House was designed and made by Juicy Glass (0115 9112715, www.juicyglass.com).

The windows were supplied by Swedish Timber Products (01347 825 610, www.swedishtimberproducts.co.uk).

The roof windows were supplied by VELUX Company (0870 405 7700, www.velux.co.uk).

For daylight pipes, try Sunpipes by Monodraught (01494 897700, www.sunpipe.co.uk).

### Design detail
Garden suncatchers by Sarah Hayhoe (www.gardenglass.net).

### Publications
*Daylighting Design in Architecture: Making the most of a natural resource* (Energy Efficiency Best Practice Programme ADH011). Free to download from the Carbon Trust (www.thecarbontrust.co.uk).

*Solar Energy and Housing Design* (S Yannas, Architectural Association 1994). A detailed guide to the many ways houses can benefit from the light and heat of the sun.

# 06 JUL 05

As our carefully crafted walls began to take shape, work began nearby on another feat of bespoke carpentry: our kitchen. This was no bad thing as we are keen cooks and didn't want the design of the kitchen to be crowded out by the many other demands of the project. In fact we had been planning the kitchen for as long as we had been designing and building the house.

The kitchen was built by Stephen Edwards in a workshop in Brixton, ten minutes walk from our building site on the Brixton-Clapham border. It was a pleasure to move between the two, first discussing the engineering of the timber frame with Steve the foreman, then the precise specification of our kitchen drawer handles with Steve the cabinet-maker.

*Genc forms the curved framework for the roof of the stair tower.*

*George begins the preliminary sheathing, creating interior spaces for the first time.*

# The kitchen

Paint and potatoes, granite and greens: these days, we care as much about how our kitchens look as we do about culinary success within them. Having emerged from dank basements and desultory back corners, kitchens have triumphed in the centre of our living spaces. For the ecologically-minded, this exposure is no bad thing, for the impact of a kitchen is a microcosm of the impact of the house as a whole: materials, energy, water and waste are core concerns of both. Make the effort to specify a stylish eco-kitchen and it will be that much easier for an eco-house to follow.

The kitchen for Tree House began life this week in a Brixton workshop where eco-designer Stephen Edwards is building cabinets from beech and birch ply, all supplied with FSC certification which ensures the sustainable management of the forests where the timber was felled. For an equally robust work surface, we have specified Kirkstone slate from Cumbria, a beautiful, durable material with low energy and environmental costs.

Lots of energy gets used in kitchens, so it's no surprise that kitchen appliances have long been the focus of energy-efficiency campaigns. The familiar A to G labels have been effective in driving improvements in design, so there are now plenty of A-rated models on the market. For the best performance, look for an A+ rating or the energy saving logo. As with many products, quality in environmental design typically goes hand in hand with quality in overall design – our AEG-Electrolux appliances will be quiet and reliable as well as energy- and water-efficient.

*The kitchen is fully integrated in the open-plan living room on the ground floor.*

To radically reduce the environmental impacts of your kitchen, you have to pay as much attention to the resources that flow through it as you do to its fabric and technology. An ultra-efficient fridge is of little value if it is filled with New Zealand apples and FSC beech is a poor disguise for bin-loads of waste heading for landfill.

To reduce the transport and packaging impacts of what comes into the kitchen, we grow fruit and vegetables in our allotment, shop at local farmers' markets and avoid buying imported food that can be produced in Britain. Design can even make a difference here: at the centre of our kitchen a wall of shallow shelves will display row upon row of glass jars containing ingredients which we can buy in bulk or from market stalls where paper bags replace plastic wrap. By using attractive Bodum jars with sustainably sourced wooden lids, this glittering and colourful expression of culinary possibility is also designed to keep us cooking and away from the temptations of take-aways and their little mountains of waste.

One of our kitchen design priorities is to treat the resources leaving it with as much respect as the resources entering it. By designating plenty of cupboard space for recycle bins we will be able to separate organics, plastics, glass and metals without getting fed up with quick-fill bins – an inevitable problem when a cramped bin space is divided into even smaller compartments. As a lot of kitchen waste is organic, including cardboard and paper packaging, the wormery in our back garden is an integral part of the design.

Although our first meal in Tree House is still some way off, plenty of food is already being produced, consumed and recycled on our site. Our great sycamore obtains raw ingredients from its immediate environment, creates food efficiently with renewable energy and ensures on-going value for its waste products as nutrition for the insect life beneath it. Our culinary ambitions may extend beyond maple syrup (sycamores are maples) but on every other count our holistic eco-kitchen will be struggling to match this towering example of sustainable design.

## profile

### Stephen Edwards

Commitment to ecological principles, attention to detail, good design ideas and genuine interest in his clients' suggestions – everything critical for our kitchen design we found in local cabinet-maker Stephen Edwards.

Ecological design has always been at the heart of Stephen's work, so we knew he had the experience to source high quality materials with the best possible provenance. Although our handmade kitchen cost more than the usual commercial alternatives, it will outlast all of them by decades.

*The kitchen incorporates energy-efficient appliances and lighting, low-impact materials and care for the resources flowing through it.*

## design detail | Cork forest

Wine producers have recently started replacing their corks with plastic alternatives in order to reduce the risk of wine contamination. The result may be a marginal improvement in wine quality but there is a larger cost. Cork-harvesting is a highly sustainable industry as the stripping of the bark from *Quercus suber*, the cork oak, does not harm the tree but encourages new growth.

Why replace the sustainable management of cork forests with a billion pieces of non-biodegradable plastic for the sake of a few unspoiled bottles of wine? Does this really amount to a net improvement in the quality of life on planet Earth? At the centre of the main living space in Tree House there is a delicate collection of ceramic and metal pots by artist Grace Collins. Its intricacy, density and changeability (it gets remade every time we dust) capture something of the landscape that inspired it: an Iberian cork forest.

## resources

The kitchen for Tree House was designed and built by Stephen Edwards (020 7737 8110, www.ecointeriors-uk.com).

The kitchen worktop was supplied by Kirkstone (01539 433296, www.kirkstone.com).

The kitchen appliances were supplied by AEG-Electrolux (08705 350 350, www.aeg-electrolux.co.uk).

The JARRA storage jars were supplied by Bodum (01604 595650, www.bodum.co.uk).

The Energy Saving Trust maintains a database of energy-efficient appliances. These can be identified in shops by the 'Energy Saving Recommended' logo (0845 727 7200, www.est.org.uk).

To find a farmers' market near you, contact the National Association of Farmers' Markets (0845 230 2150, www.farmersmarkets.net).

Local recycling banks can be located through www.recycle-more.co.uk.

Publication: *Composting with Worms: Why waste your waste?* (G Pilkington, Eco-logic Books 2005). Includes a guide to making your own wormery.

# 20 JUL 05

I was inspired to write this diary entry after a visit to our friends Karen and Pete, shortly after Karen had given birth to her first baby, Madeleine. I did contemplate asking them to come over for a photocall in the salvaged bath in our back yard but Madeleine seemed a little too new to the world for such capers.

Back on site, the completion of the sheathing at the front of the house gave us our first impression of the solid form of the building, albeit without its all-important roof. We could now stand in the space that would become our bathroom and imagine every final detail in place.

# Baths and showers

Has anyone ever thrown the baby out with the bathwater? This event seems to me to be so unlikely that even its existence as a figure of speech seems inexplicable. Did bath design go through a bad patch at some point with calamitous consequences?

Although a tad surreal, the expression seems to capture our nervousness about the risks of radical action. For our Clapham self-build, the proverbial bathwater consists of all the conventional assumptions about how houses should work, the established compromises about carbon emissions, resources, pollution and waste.

*Steve and Genc screw the plywood sheathing on to the curved stair tower.*

If it is possible to generate all your own energy, use entirely non-toxic materials and send next to nothing to landfill, why do otherwise? However, given public suspicions of what this involves in practice, we are also very focussed on our baby: domestic bliss with all mod cons. We are confident that contemporary environmental design not only holds on to this baby but significantly improves its health. Ford and I plan to enjoy life in

Tree House without any sacrifice in quality of life and all the benefits of a warm, light-filled, healthy home.

Which brings me to the important issue of our bath. London is the most water-stressed city in Europe: the south-east has suffered a series of winter droughts in recent years, and the Environment Agency is urging us all to take showers instead of baths. Under these circumstances, you might expect me to have removed the bath from our radical environmental specification. Except, of course, this would inevitably involve throwing our quality-of-life baby out with the eco-compromised bathwater.

The key issue is not so much owning a bath but how often you use it. For us, baths are not instruments of cleanliness but of relaxation, designed to meet a very particular need after (say) a long day digging the allotment. As this need is relatively infrequent, we take few baths. But when the need does arise, nothing can replace a good soak.

*The walk-in shower is no less attractive for being the water-efficient choice.*

A typical bath uses 80 litres of water compared with only 30 litres for a shower, so it makes good sense to opt for a shower over a bath for ordinary ablutions. However, showers are more convenient to use than baths, as well as more water-efficient, so if you currently live without a shower, don't rush to install one if your twice weekly bath will be replaced by daily showers. Above all, if you are planning on installing or improving your shower, don't be tempted by a power shower – an environmental design disaster that uses more water than a bath for a marginal gain in physical sensation. Instead choose a shower with a low-flow option on the showerhead and responsive controls that encourage quick stop-start actions: stopping to lather up, restarting to rinse off.

Having made it on to the specification, the bath for Tree House (currently in our back yard) comes with excellent eco-credentials in every other way. Firstly, it's salvaged: an old roll-top bath I bought at auction for a knock-down price. Second, the original surface is intact, so the toxic and highly energy-intensive process of

*The solid form of the house, sans roof, emerges.*

re-enamelling is not required. Third, we are replacing the rusted taps with a new pair designed to fill the bath rapidly and so minimise heat loss before you get in. This is the opposite strategy to all the other taps in the house which have built-in flow-restrictors and aerators to give the best results for the least water throughput (Hansgrohe taps include these as standard).

To really save water, there's always the self-cleaning option. Not easy for humans, but a highly effective feline strategy. As our four cats are not only self-cleaning but also integral to our quality of life, there's absolutely no way, real or proverbial, that these babies will ever get thrown out with the bathwater.

*The bath, looking on to a stunning wall of Kirkstone Sea Green slate, is designed for the occasional indulgence, not regular washing.*

design detail

## Laundry sink

Our little laundry boasts a salvaged Belfast sink set on a birch table made by our kitchen designer, Stephen Edwards. We also have a washing machine in the room but there are plenty of occasions when a spin in the sink seems like a better idea. We deliberately chose a big sink in order to make hand-washing an attractive option. After all, who wants to soak their sheets in a sink designed for teacups?

resources

The water-efficient taps and showers for Tree House were supplied by Hansgrohe (0870 7701972, www.hansgrohe.co.uk).

### Design detail
Belfast sink supplied by Lassco (020 7394 2101, www.lassco.co.uk).

### Publication
*Conserving Water in Buildings*. Eleven concise fact sheets on water efficiency in homes, available from the Environment Agency (08708 506 506, www.environment-agency.gov.uk).

## *27 JUL 05*

**The delay in the construction of our roof, due to the lack of materials, was brief compared to our earlier problems but it did get under my skin rather badly.**

**Months later my contractor told me that far from neglecting the build, he had been working hard to get a good price for the materials. It's a great pity he felt unable to communicate this at the time, for I was left feeling so powerless that I briefly felt myself despising the whole project. A passing moment, thankfully, but a nasty one.**

# The Association for Environment Conscious Building

Drop the words 'self-build' into conversation when you're next down the pub and I can guarantee that within a mere sixty seconds the words 'Kevin McCleod' will also have been uttered. Kevin is to self-build what Queen Victoria was to the British Empire: patron and final arbiter of the dream-building elite, a significant achievement that has undoubtedly helped the great British Channel Four-watching public to think about their living spaces with greater imagination and ambition.

*Friends Katie and Simon provide some welcome reassurance about the rewards awaiting us at the end of the project.*

Tree House was not considered to be grand enough for Grand Designs, so the inevitable question in the pub does become a bit irritating after a while. In retrospect, however, we are deeply relieved not to have the extra stress of co-ordinating a television crew within a schedule that is frankly a complete fantasy, even on a day-to-day level. Bursting into tears on Kevin's shoulder on prime-time TV would also have been less than totally cool.

Although I'm always upbeat in this column, sustained by my undimmed enthusiasm for eco-design, I have to confess that the last ten days have been pretty grim. Two weeks ago, our timber frame finally reached the crucial moment for the roof to go on. This point in the project had been anticipated for some time but, despite foreman Steve's constant reminders, our contractor did not get round to ordering the materials for the roof until the day that they were actually needed on site. Two weeks later, we still have no materials and no roof. It has become all too clear that for all his attention to architectural detail, our contractor's sense of timing will never get him a job on the comedy circuit.

*Unlike the building, the garden is on schedule. Brother Hugh makes his judgement.*

All professional-client relationships have their sticky moments and all clients need to run away now and then. I am personally feeling much better after taking the waters at the annual conference of the Association for Environment Conscious Building (AECB) in Taunton. The conference took place at the Somerset College of Arts and Technology where the Genesis Centre is currently being built, an exciting building that will demonstrate in its own fabric a wide variety of approaches to sustainable construction. As well as exploring such issues as building with cob, sourcing reclaimed materials and the very best practice in eco-renovation, I was also able to have a good gas with lots of other like-minded people about the highs and lows of eco-building. Contractor problems? So what's new under the sun?

The AECB is a stimulating network of people committed to changing – radically – the way we build. It has a large professional membership but welcomes self-builders and eco-building enthusiasts of all kinds. The association website gives details of how you can join, membership benefits and a range of information about environmental issues in construction. As the AECB has a strong lobbying focus, this is a good network to join for the nation's benefit as well as your own.

I have found that another way of getting some much-needed perspective on our problems is to show people round the site, preferably friends who have not seen it for some time. This always leads to appropriate cooing and reassurances that the house will indeed be absolutely fabulous, although I think our friends Katie and Simon rather missed the point in admiring the 'sun trap' on our roofless top floor.

One way or another, my optimism and enthusiasm always return. But there is a chill at my back. I really didn't think I would ever have to face this, but for the first time I have to take seriously the dreaded question pressed upon me by the ghost of Kevin McCleod: 'But will you be in by Christmas?' All I can say, in truth, is that we have absolutely no idea.

## *design detail*  Stair tower roof

This little roof did not have the power to spoil any working relationships. It protects the stair tower – the curving trunk at the front of the house – and is clad in cedar shingles to tie in with the cedar cladding of the tower.

Shingles are tiles made from timber and, like ordinary tiles, can be used as a rainscreen for walls as well as for roofs. As they have to cope with serious weather, they are made from woods that are naturally durable. Cedar shingles have quite enough resins packed in them to keep the rain off for decades without the need for chemical treatment.

## *profile*  Genc Marku

Genc joined the team when the frame was going up and stayed through to the end – Steve always ensured that the most skilled workers stayed with the project. Genc was a great carpenter but was happy to turn his hand to other jobs when required. The only task he did not seem to enjoy was deputising for Steve when he went on holiday (too much grief by a long way).

## *resources*

Association for Environment Conscious Building: www.aecb.net. Sign up for a regular newsletter, the excellent *Building for a Future* magazine and a lively annual conference.

To find professionals who are members of the AECB see the website or consult *The Green Building Bible* (K Hall, Green Building Press 2005, www.newbuilder.co.uk).

The Genesis Centre: www.genesisproject.com. As well as demonstrating different approaches to ecobuilding in its fabric, the Centre offers short courses and runs events.

# *03 AUG 05*

The materials for the roof arrived at the beginning of August, at which point the complexity of the design became fully apparent. Although our architect had unquestionably excelled himself, his elaborate structure of exposed trusses was not going to be quick to build. As there were a lot of crooked angles to cut in rather expensive wood, I was very glad that the guys on site had all shown themselves to be skilled joiners.

Other jobs also proceeded, not least the layer-by-layer build up of the wall structure. Walls are surprisingly complicated bits of the building fabric as they play such a key role in moderating the climatic interactions between the inside and outside of the building. Our approach to tackling the problems of heat loss, airtightness, ventilation and water vapour is one among many.

*The trusses are made from honey-coloured Douglas fir and ordinary softwood.*

*Steve and Genc begin cutting the complex roof trusses to size and shape.*

## Water vapour

Good God, I'm picking colours. After ten months of devotion to foundations and structure, could this be the dawning of the era of colour charts and soft furnishings?

Actually, this era dawned a long time ago – it's no good leaving such things to the last minute. But it will be some time before interior design becomes the focus of the Tree House building site. The colour I'm picking is for the render board that will provide much of the 'rainscreen' for the building, complementing the timber-clad trunk that wraps around our staircase. Both types of cladding will soon be nailed in place on timber battens, leaving a ventilation gap between the rainscreen and the main wall.

Our walls must obviously be robust enough to stop water getting in but this is a simple task compared to dealing with moisture going the other way. The principal movement of moisture within walls is from the warm, humid interior to the cooler, drier exterior. If your brick walls weep, this is why.

*The tree casts its morning shadow over the bright white render board rain screen at the front of Tree House.*

We generate huge amounts of water vapour in our homes: cooking, bathing, washing and breathing all increase humidity. This can cause all sorts of problems, not least asthma: 80% of asthmatics are allergic to the droppings of dust mites, which proliferate in warm, humid conditions. For every wisp of steam that you see rising from your coffee cup, potato pan or bathtub, there's a whole lot more that you can't see. In fact, what you see over your coffee is not vapour at all but the condensation where the rising gas meets cooler air.

Condensation occurs whenever air cannot hold any more water. This happens in hot conditions if there is lots of moisture in the air, such as in a steam room or a tropical city in the wet season. But condensation is much more common in Britain in cool conditions because cold air can hold very little moisture. Cold air is very dry and may not even absorb our rapidly condensing winter breath. In houses, condensation is common in poorly insulated cold spots such as window frames where warm air meets the cold surface. However it is a greater problem if it occurs in areas you can't see, such as inside your walls or roof, where it can do everything from undermining your insulation to rotting your building fabric.

*The Tyvek vapour-permeable air-barrier wraps around the building.*

There are three ways of dealing with water vapour: a) get rid of it with good ventilation; b) put a barrier in your wall to stop it getting in; or c) design your wall to cope with it. The most advanced version of (c) is a 'breathing wall' which absorbs the water vapour produced in the building and carries it outside without the use of any active ventilation systems. This is the method used by the two billion people who live in houses made of earth and clay.

Our approach is to pursue all these strategies at once with an eye for energy conservation. For ventilation, we will open the windows in summer but turn on the mechanical ventilation in winter: our Vent-Axia 'LoWatt' whole-house ventilation system combines low power consumption with a very high rate of heat transfer from the outgoing to the incoming air. To stop the vapour getting into the walls, the entire shell of the building has an internal lining of recycled black plastic that also helps to stop valuable warm air escaping. Finally, if moisture does get into the walls, the Warmcel recycled newspaper insulation will carry it out to the ventilation gap behind the rainscreen, preventing any moisture build up within the timber frame. The result will be a house with high air quality, low humidity and very low energy losses.

As for that colour choice, I confess I didn't even look at the lovely charts in the catalogue. According to the Bauhaus school, white is the colour to use if you want to draw attention to the form of a building rather than its finish. Although the tree-like form of the house may not be entirely to the modernists' taste, their advice still holds. A bright white finish will emphasise the curvaceous form and timber details of the front of Tree House, letting our arboreal intent shine through.

*design detail*

## Buoy

Of all the materials and objects within Tree House, this one has been exposed to by far the most moisture and still survived almost intact. We found it on a beach on Feochag Bay in north-west Scotland, long abandoned and washed by innumerable tides.

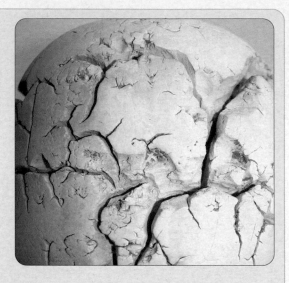

It is one of our most beautiful possessions. It has the colouring and delicacy of a ceramic vase combined with the character of a material that is old, worn and burnished.

The buoy has inspired an ongoing pursuit of found objects. The progress of this hobby remains slow but it does make you look at everything twice.

*resources*

The whole-house heat recovery ventilation system for Tree House was supplied by Vent-Axia (01293 530 202, www.vent-axia.com).

The Warmcel 500 insulation was supplied by Excel Industries (01685 845200, www.excelfibre.com).

Gaia Architects (www.gaiagroup.org) have been exploring the design of low-allergen housing using benign materials and careful ventilation, particularly at Fairfield Toll in Perthshire.

### Publication
*Energy efficient ventilation in housing* (Energy Efficiency Best Practice Programme GPG268). Free to download or order from the Energy Saving Trust (0800 915 7722).

# 10 AUG 05

It took time to cut all the roof components to their precise size and shape but, once this was done, the primary structure went up fairly quickly. Not that this was an easy job: it involved manoeuvring odd-shaped trusses with bespoke wall and floor junctions onto a wall frame that in one corner rose sharply and at an angle, creating yet more challenges for the carpenters.

Elsewhere, in Brixton, our kitchen was taking shape. It was obvious that the kitchen would be complete long before the house was ready for it but this was a relatively small problem in the scheme of things, even when the component parts of the kitchen occupied the front room of our rented flat for four months.

# Induction hobs

What did Michael Faraday – the man on the twenty-pound note – cook his dinner on? Although the coal-fired range did not advance very far in the course of Faraday's life (1791-1867), his relationship to this technology must have changed profoundly. Born the son of a blacksmith, he would have gained an early knowledge of the fire and smoke of hot forges and hot stoves alike. Seventy years later, as an esteemed member of London society, he had the means to leave the sweat of the Victorian kitchen to others.

*The induction hob sits neatly in the slate worktop of the completed kitchen.*

Such socio-economic transformation remains the most reliable non-cooperative way of escaping from environmental pollution. The distinction between upstairs and downstairs in the Victorian household is reflected in the class distinctions of western and eastern districts of British cities (the latter are downwind of the city's pollution). Today the West's voracious consumption of raw materials is blind to the environmental exploitation of poorly regulated, non-unionised labour in developing countries where the materials are produced.

Faraday did not make his name tackling such environmental injustice but in developing a technological, rather than socio-economic, means of escaping the effects of environmental pollution. In 1831 he demonstrated the principle of electro-magnetic induction: the creation of electric current in a conductor by a moving or changing magnetic field. This discovery meant that electricity could leap from the realm of scientific curiosity into the brave new world of power technology: generators, transformers and innumerable electric devices began life here.

Electricity is magical. Not only is it incredibly versatile, it is also pollution-free at the point of use and you don't have to belong to the upper classes to enjoy its smoke-free benefits. Unfortunately, however, this magic is a trick. Of all the different forms of energy we use in our homes, electricity is the dirtiest because over half the energy released from the coal, gas or oil burnt in the power station (to drive the turbines that drive the generators) is wasted. The hidden injustice here is global, as the worst impacts of climate change, fuelled by the power station's carbon dioxide exhaust, will be borne by countries with the least means to cope with them.

The inefficiency of traditional electricity generation means that it is usually better to cook with gas than electricity. Unless, that is, you take Faraday's discovery one step further and install an induction hob. An ordinary hob works by emitting heat upwards to the base of the pan; an induction hob stays cool but creates heat within the pan by way of Faraday's changing magnetic field. The result is almost a doubling of energy efficiency from around 45% for an ordinary gas or electric hob to 82% for an induction hob.

As all the energy for Tree House will be generated on our solar roof we don't have to worry too much about fossil fuels and distant power stations. Nonetheless, we can only achieve our energy self-sufficient goal by radically reducing our demand for energy in the house. Given the considerable energy needed for cooking, our AEG-Electrolux induction hob is a godsend and already has pride of place in the kitchen taking shape in Stephen Edward's Brixton workshop. Happily, as well as being incredibly efficient, induction hobs are also highly responsive, very safe and easy to clean.

Although I'm a sucker for technology, I couldn't end this piece without suggesting a little behaviour change to improve your cooking efficiency: keep lids on pans, turn the ring off and let the pan do the last five minutes by itself (unless you are frying) and, if you cook with gas, make sure the pan covers the flame.

*The first branches of Tree House emerge.*

Alternatively, if you have already made it to the top of the socio-economic pile, I recommend that you dispense with a hob altogether and live entirely on rocket salad, sashimi and steak tartare.

## Hub dynamo

Electromagnetic induction is brilliantly exploited in this hub dynamo on our Brompton folding bicycles. The dynamo is integrated into the hub and makes the most of the rotation at the centre of the wheel to create electricity to power the lights on the bicycle. A hub dynamo is more efficient than tyre-drive dynamos and does not cause drag. It is also more robust.

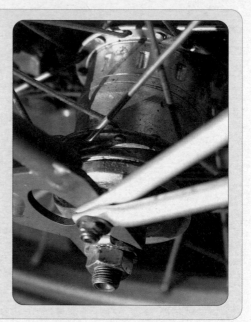

Kitchen design and build by Stephen Edwards (020 7737 8110, www.ecointeriors-uk.com).

Kitchen appliances supplied by AEG-Electrolux (08705 350 350, www.aeg-electrolux.co.uk).

The Energy Saving Trust maintains a database of energy-efficient appliances. These can be identified in shops by the 'Energy Saving Recommended' logo (0845 727 7200, www.est.org.uk).

Useful information on low-energy cooking can be found at The Yellow House (www.theyellowhouse.org.uk).

Son dynamo on titanium Brompton bicycle (020 8232 8484, www.bromptonbicycle.co.uk )

# 17 AUG 05

As August progressed, our roof blossomed. Our architect had modelled our top room using a computer graphics package so we had a good idea of what the experience of being in it ought to be like. Nonetheless there is always a risk that such an ambitious and unusual space does not turn out quite as expected. Thankfully there were no disappointments in store.

Topping out is always an important moment in the life of a construction project and ours was no different. Not everyone on site was convinced (yet) of the merits of the canopy of Tree House but nobody was looking for work elsewhere – this was now definitely a job worth seeing through.

*The architect's medieval sensibilities become increasingly apparent.*

## Heartwood and sapwood

It's the height of summer and the canopy of Tree House is finally filling out. Great branches of Douglas fir unfurl in the big open-plan study at the top of the house, ready to lift our long-awaited roof into place. It feels as if the flourishing tree at the front of our site and the house that ever so slightly wraps around it have begun a long and intimate embrace.

Reflecting on our guiding metaphor, I am struck by how freely we draw on the natural world to illuminate our own, yet rarely return the favour with any sympathy, perhaps because projections of our manufactured world on to the natural landscape tend to expose uncomfortable truths about the relationship between the two. At best the biosphere is a garden, more often a playground, warehouse or rubbish dump.

The world would be a happier place if our first intuition was to treat the natural world as a university, a place to seek wisdom, skills and technology. This approach is most fully articulated by the advocates of 'biomimicry', who point out that many of our energy-intensive, pollution-saturated technical achievements have long been realised in the natural world at ambient temperatures with minimal resources and entirely beneficial side effects. The lotus stays remarkably clean in muddy Asian swamps because dirt particles cannot hold to the microscopic mountain ranges on the surface of its leaves. A spider excretes a material for web-building that, weight for weight, is five times stronger than steel. The abalone shell is one of the hardest materials known to man yet it is made in the sea rather than in a furnace.

Trees are fine examples of biological engineering, but our efforts in Clapham to build a house that 'works like a tree' are focussed on achieving the ecological performance of a tree without necessarily using arboreal technology. For example, a tree harvests solar energy using tiny organic photosynthetic solar cells, largely made from air and water, whereas our photovoltaic

*George, Pete, Steve and Genc admire their work.*

solar roof is made from highly refined materials, leaving a significant trail of energy emissions and waste behind.

The nearest we will get to working literally like a tree is in our extensive use of wood. When you chop a tree down, the transport of water and sugar within the wood ends, but the extraordinary strength, resilience and durability that sustained the tree remain. Has mankind ever built such an elaborate, towering structure on a single slender post, fit to last hundreds of years?

In fact, the best timber comes from the inside of the tree where the wood is effectively already dead. As a tree grows and its trunk thickens, the tree packs oils and resins into its core. This heartwood protects and strengthens the tree and helps to support the sapwood around it where the living processes continue.

Although most timber sold commercially is heartwood, it sometimes comes with edges of sapwood. These are vulnerable to insect attack but the heartwood will only get munched if it has already decayed, usually because of prolonged exposure to damp. The sculpted look of timber beams in old buildings is due to the sapwood edges being eaten, leaving the resilient heartwood behind.

So don't rush to spray your home with nasty pesticides if you see tiny holes in exposed timbers as this is likely to be expendable sapwood. Such 'flight holes' are usually historical anyway: your guests may have left decades ago, especially if you have central heating that dries out the wood and makes it unpalatable. Where wood treatment is unavoidable, use boron-based products which have low toxicity to both humans and the environment.

Foreman Steve is not entirely persuaded by our branching roof, as not all of the branches are strictly necessary. Happily, though, we are not inspired by hard-line modernism but by a tree, and in the natural world lots of things are not strictly necessary: abundance is the reward of truly sustainable design.

*profile*  Pete Shovlin

Despite his name, Pete managed to avoid the long months of muck-shifting, joining the team when the woodwork was getting interesting. His father worked a peat bog but Pete was clearly happier up in the rafters. A great joiner in the making.

## design detail

## Salvaged oak window surrounds

The sharp white exterior walls of Tree House are very contemporary but they are complemented by a deliberately ancient detail. The window surrounds on the front and rear of the building are made from oak beams salvaged from a barn restoration in Sussex. The beams have seriously rough edges, reflecting long-gone insect attack and the fashioning of earlier carpenters, whose marks can still be seen on the timbers.

These chunky frames, clasping the windows, play with the idea of light seen through the broad branches of a tree. They also physically link Tree House to the long tradition of timber building in Britain.

## resources

The book that raised the profile of biomimicry has the same name: *Biomimicry* (J Benyus, William Morrow 1997). A good overview is offered by www.biomimicry.org.

### Publications
*The Green Building Handbook* (T Wooley, S Kimmins, P Harrison and R Harrison, E & FN Spon 1997). Includes a detailed chapter on timber preservatives.

A *Pests in the home* factsheet on timber treatments is available from the Pesticides Action Network (www.pan-uk.org).

# 24 AUG 05

From the outset of the project, my ambition to build a 'zero carbon' house had included a determination to demonstrate how (and whether) we achieved this, and in some detail. I was amazed to find that there were no commercially available products on the market that could give me accurate and detailed information about the energy we generated and consumed within the house.

So I decided to develop such a system. Fortuitously, I met Luke Nicholson and Ben Pirt of More Associates at just the right time: they had already developed exciting energy feedback systems in the commercial sector and were keen to explore the possibilities in homes. With the support of Powergen, an ambitious system for recording and communicating every power event in Tree House was born.

Back on site, the roof's plywood sheathing was fitted over the lovingly prepared trusses. This was the first layer of the big south-facing solar power station that defines the form of the building.

## Energy feedback

What's a watt? A what? A watt! – what about a kilowatt? A kilo-what? Oh come on – how about a kilowatt-hour? Isn't that something on my electricity bill? Yes! – but what?

*Luke Nicholson shows off his 'sunfall' display showing solar power generation.*

Under the sheathing, the
VELUX windows go in.

Nik and Genc install the roof's
plywood sheathing.

Although our contemporary world is increasingly obsessed with energy supply and energy efficiency, our collective grasp of exactly what is going on in our domestic wires remains extraordinarily weak. Can you describe a kilowatt-hour and its relationship to all the lights and gizmos you turn on and off in your home?

Our confusion about electricity consumption could easily be overcome with the help of some clever low-powered technology. Imagine if you had a discreet control panel in your kitchen where you could find out how much electricity every appliance and light circuit in the house was consuming at that moment, complemented by tiny ambient indicators in every major room (perhaps above the light switches) that changed colour according to the overall energy consumption of the house. Imagine if all this information was digitally recorded and turned

The prototype
energy feedback
display
developed for
Tree House by
More Associates.

into a computer-based analysis that made very clear, in units of your choice (cups of tea?), exactly how your house performed over the weeks and months and what differences your choices within it actually made. There is already good research evidence to show that such feedback could lead to cuts in power consumption of between 10 and 20 per cent.

Although such a product does not currently exist, this may be about to change. With the support of Powergen, I have just begun work with Luke Nicholson and Ben Pirt of More Associates to develop a system for Tree House to the specification described above. Their experience of designing user-friendly energy feedback systems in commercial environments should be invaluable in tackling the relatively uncharted territory of the British living room.

It might seem a little odd to develop such a system for a house that will be ultra-efficient from day one, but we hope to use the house as a test-bed for developing energy feedback products that will be commercially viable. The energy map of Tree House will also be available online, enabling any web users not only to find out what we are up to, but also to match their own lights and appliances with ours and see what difference this makes to the economic and environmental bottom line.

You might also think it odd that a power company would want to invest in a means of reducing demand for its own product, but all electricity suppliers are required to do this by law. This 'Energy Efficiency Commitment' is achieved by insulating homes and encouraging the uptake of more efficient lighting, heating and appliances.

## *design detail*

## Integrated digital television

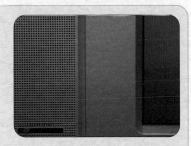

There is one existing energy feedback mechanism that proliferates throughout our homes: the little red stand-by light on electronic equipment of all kinds. Unfortunately this light all too often gets ignored, with the result that we need a couple of power stations just to keep all this kit humming in the background doing nothing. The more boxes we have with little red lights, the less likely we are to turn them all off. Box integration is therefore good environmental design. As our LCD television has a built-in digital decoder, we do not need a separate digibox and only have one button to press to turn the TV off properly. This may be a small energy-saving detail but when multiplied by a population of 60 million, such details make a difference.

How are you faring with those watts and kilowatt-hours? The problem here is the lack of a clear unit of energy. Usually, we begin with such a unit – gallons, calories, donuts – and then define a rate against it: miles per gallon, calories per meal, donuts per day. But for electricity, we begin with a rate of consumption, the watt, and then create a clumsy unit of energy with it: one kilowatt-hour of energy is expended by a kilowatt (1,000 watt) fire in one hour, or by a 100 watt bulb in ten hours.

It makes more sense to begin by focussing on the basic unit of energy that the watt itself describes: the joule (named after an English brewer not a French scientist). One watt is a rate of energy consumption of one joule per second (J/s). So one kilowatt-hour of energy is actually 1,000 x 1(J/s) x 3,600 (seconds in an hour), which equals 3,600,000 joules or 3.6 megajoules. Perhaps if we all worked in scary megajoules rather than wimpy kilowatt-hours, we might take our power consumption a little more seriously. Anyone fancy a pint?

*profile*  ## Nikolin Deda

Nik joined the team when the house was reaching its structural peak and helped to ensure that the quality of the build remained at its highest for the rest of the project. With a background as a cabinet-maker, Nik had an enviable eye for detail. Whenever adjustments had to be made to the architect's drawings to cope with the reality on site, Nik always ensured that the end result was an improvement on the drawings.

*resources*

The energy monitoring and feedback system for Tree House was designed by More Associates (020 7841 8904 www.moreassociates.com) and funded by Powergen.

For information on the energy efficiency services that Powergen offers to everyone, contact the Powergen energy efficiency line (0500 20 10 00, www.energyefficiency.powergen.co.uk).

The Energy Saving Trust offers advice on reducing power consumption in the home (0845 727 7200, www.est.org.uk).

# 07 SEP 05

Insulation is at the heart of eco-building, but it took me a year to get round to writing about it in any detail. This was principally because the first insulation we fitted was on top of the roof. We then worked downwards, filling the walls and then finally laying the floor insulation on the slab, over which the underfloor heating was installed.

It was exciting up at the top of the scaffolding watching the roof take shape. Once the main structure was in place, it moved quickly, the layers of plywood and insulation giving the roof a much-needed sense of solidity. The views of London in the red haze of late summer dusk were also fabulous.

*George fits the long strips of insulation between battens on the roof.*

*Pete cuts the boards of insulation into strips.*

## Insulation

Do you count sheep to help you get to sleep at night? On cold nights, do you dream of sheep? Have you ever dreamt of a stoned sheep in a cork forest reading a copy of the *Daily Mail*? You have? If so, turn the page immediately because you have already spent far too much time obsessing about insulation.

I've been dreaming of insulation for months, but last week we finally fitted the first pieces into Tree House. Grabbing the opportunity of a dry day between the storms, all hands were on deck – the roof deck – to build the thick layer of insulation between the roof's plywood sheathing and the plastic rain screen that will support our solar power station. By the end of the day the job was done and the roof was covered in a bright blue tarpaulin just in time for the next downpour.

Our super-insulated roof is not graced with any of the materials alluded to in my dream: hemp, sheep's wool, cork and recycled newspapers. These are all excellent insulants, but although cork (the bark of the oak *Quercus suber*) would perhaps be the most fitting choice for Tree House, our roof and floor are in fact being packed with a synthetic product, Kingspan Insulation rigid phenolic boards.

The key issue with all insulation is to select the most appropriate material for the task. Bark may be all that a tree needs to protect its living tissues from the extremes of the elements but our 'zero-carbon' house needs something more substantial. We have chosen rigid phenolic boards for the roof simply because they keep the heat in (or out) better than any other material on the market.

The most important characteristic of any insulation is its thermal conductivity, so always check the value of this when making comparisons. The thermal conductivity of a material is the rate at which heat flows through it when the temperature on one side is greater than the other. This is standardised as the rate of energy flow (in watts) through one metre of material for one degree difference between inside and outside (in degrees Kelvin – the same scale as Centigrade, but starting at absolute zero). Typical values are 0.034 W/mK for mineral wool, 0.037 W/mK for sheep's wool and 0.022 W/mK for phenolic board. The lower the conductivity, the more effective the insulation.

Beyond this bottom line, there are lots of other issues to consider in choosing insulation such as the size and shape of the space, moisture and fire risks, and the potential for poor installation to undermine effectiveness. For example, we will be spraying Warmcel 500 recycled newspaper into our walls because the nooks and crannies of our 'I beam' wall studs would not be adequately filled with rigid boards.

Synthetic insulation materials have often been dismissed by eco-builders because of the blowing agents used in their manufacture. These were originally ozone-depleting CFCs (chlorofluorocarbons), subsequently replaced by marginally less nasty HCFCs (hydrochlorofluorocarbons) and HFCs (hydrofluorocarbons). Happily, Kingspan Insulation now produces all its insulation using pentane which has minimal impact on either the ozone layer or global warming, and any energy used in manufacture is rapidly offset by its energy-saving performance once installed. In years to come, our super-insulated roof will keep us cool in hot summers as well as warm in cold winters, so we should sleep soundly beneath it. With any luck, my dreams will be more peaceful too, with all those giggling right-wing sheep long chased into the wilderness.

## design detail — Bed cover

The personal insulation we use in our daily lives is just as important to our thermal well-being as the stuff we wrap our houses in.

Our tree-like ambitions have inspired our very good friend Sara Maitland to embark upon an extraordinary labour of recycling with the ultimate aim of improving our personal insulation. She has been collecting left-over materials from near and far and is now sewing a patchwork of trees that will become a full-sized bed cover. Her work is in the tradition of the Amish who make bed covers out of the excised pockets of denim shirts, though quite why these pockets are a danger to the spiritual life we do not know.

## resources

The phenolic board insulation in the roof and floor of Tree House was supplied by Kingspan Insulation (0870 850 8555, www.kingspan.co.uk).

The Warmcel 500 recycled newspaper insulation in the walls of Tree House was supplied by Excel Industries (01685 845200, www.excelfibre.com).

For advice about the insulation in your walls, wall cavities, floors, loft or roof, contact the Energy Saving Trust (0800 915 7722, www.est.org.uk).

### Publications

The following Best Practice in Housing publications are free to download or order from the Energy Saving Trust: *Advanced insulation in housing refurbishment* (CE97), *Cavity wall insulation in existing housing* (CE16/GPG26), *Cavity wall insulation: Unlocking potential in existing dwellings* (GIL23), *Domestic energy efficiency primer* (CE101), *Effective use of insulation in dwellings* (CE23), *External insulation systems for walls of dwellings* (CE118/GPG293), *Insulation materials chart – thermal properties and environmental ratings* (CE71), *Internal wall insulation in existing housing* (CE17/GPG138).

# 14 SEP 05

During September the fragility of the building was slowly transformed by the emergence of a robust and solid exterior. The insulation-packed roof was prepared for its solar panels and the walls were clad in board and then rendered. Unfortunately the stair tower at the front of the building had to wait another three months for its timber cladding but, despite the absence of this important detail, the house was definitely beginning to burst through.

Although our solar thermal (hot water) panel was delivered in September, it was not finally plumbed into our heating system until the following January. This was partly because our plumber had to be convinced about the unusual tricks we were playing with the solar panel and heat pump. We were certainly pushing the envelope here – I had to go a remarkably long way to find an effective means of integrating these two renewable technologies.

## Solar thermal

It's a hot September day in London and the sun is beating down on the tanned backs of Team Tree House. From the top of the scaffolding we can see for miles: thousands of hot roofs under which sweaty citizens are showering, the water pouring over them heated by gas piped from deep beneath the North Sea (before long, from Russia). Some

*The gutters begin to take shape and the walls are prepared for rendering.*

people think that renewable energy is cranky; personally I think this taken-for-granted thermal landscape is stark raving bonkers.

If every south-facing roof in Britain had a solar panel, we would make a huge cut in our national carbon emissions. If you have such a roof, make the most of it: grants are available, the panels are not difficult to install and you will get plenty of hot water even on cloudy days. Unlike photovoltaic (solar electric) panels, solar thermal panels are a simple and cheap technology, a radiator in reverse on your roof.

Your solar panel will provide 60-70% of your hot water over a year. Unfortunately, they are no good for central heating as the panel rarely gets hot in winter. A solar panel can't help you if the temperature of the panel falls below the temperature of the water in your hot water tank, usually around 50°C (much less when you empty it). So there is quite a lot of the year when the panel is warm, even quite hot, but unable to do anything useful.

That is, until now. Last week I turned my back on the sun-baked south and ventured as far north as I possibly could in one day without flying: the Orkney Isles. The following morning I was greeted in the Orcadian drizzle by Alton Copland, director of Ice Energy Scotland, and a man on a mission to transform his established business – selling heat pumps – with some innovative integrated technology.

There are three deep bore-holes under Alton's house containing pipes carrying a refrigerant, which extract warmth from the ground to heat the house above. Ordinarily, the fluid comes out of the ground at around 6C, which his heat pump converts to a 50°C hot water output using simple fridge technology. The heat in the ground may be there for the taking but a lot of electrical energy is needed to shift it and compress it, typically one unit

*Calvin and Pete lay a thin layer of render on the external walls.*

163

*The roof is now almost ready for a covering of solar technology.*

of electricity for every 3–4 units of heat moved. An efficiency of 350% is pretty good but it would be considerably higher if the ground temperature was just a few degrees warmer. If only there was a free source of energy available to stick down the pipes and raise the ground temperature, ideally a warm liquid between 12°C and 50°C, as anything hotter might damage the pipes . . .

Sure enough, sitting proudly on Alton's rain-spattered roof is a big black solar panel. When the panel is hot, it heats his water tank directly. When it cools, it gets redirected into the ground, providing that extra lift for the heat pump. Because the solar panel operates at such low temperatures, its output rockets, contributing not only to hot water but also to central heating (via the heat pump) across the year. And, needless to say, Alton's energy bills tumble.

The idea is brilliant but it's taken a lot of trial and error to develop a robust commercial product, as Vera, Alton's wife, will somewhat wearily confirm. Happily, that moment has now come and Britain's third Ice Energy Solar kit (patent pending) will shortly be installed in Tree House, connecting the ground pipes we sank last November with our big bright solar panel. As we will be literally saving the sun for a rainy day, we will surely have the sanest rooftop in the whole of our crazy capital.

*The roof-top solar panel will feed the ground loop as well as the hot-water cylinder.*

## Bath taps

All the taps in Tree House
have flow regulators and
aerators to prevent excess
flow of precious hot water –
except for the bath taps. In
this instance, we want the
hot water to flow as fast as
possible in order that the
heat isn't lost before the
bath is occupied.

*resources*

The Ice Energy Solar Kit for Tree House was designed and supplied by Ice
Energy Scotland (0845 600 1020, www.iceenergyscotland.co.uk). The solar
panel was manufactured by AES (01309 676 911, www.aessolar.co.uk).

For information on the grants available for solar panels, contact the Energy
Saving Trust (0800 915 7722, www.est.org.uk).

The Solar Trade Association is a trade organisation providing news and
information about the technology (www.solartradeassociation.org.uk).

### Design detail
Traditional bath taps from Aston Matthews.

### Publications
*Solar air collectors for buildings* (Energy Efficiency Best Practice Programme
GIR59). Describes how solar collectors can also heat the air supply to a
building. Free to download from the Carbon Trust (www.thecarbontrust.co.uk).

*Tapping the Sun: A guide to solar water heating* (C Laughton, Centre for
Alternative Technology 2004). A concise guide from an experienced installer.

# 21 SEP 05

When the walls of Tree House were filled with pulped newspapers, the character of the building changed quite suddenly. Instead of rattling around a great big noisy wooden box, we were wrapped up in an enormous grey blanket: warm, quiet, even cosy.

The Warmcel 500 insulation came in compressed blocks, which were fed into a hopper and then pumped into the walls of the building. Needless to say, the stuff got absolutely everywhere – we were sweeping up grey mulch for weeks to come.

## Recycled products

Britain's recycling logo, a simple green ring with a heart shaped arrow, suggests that recycling is something that we can all learn to love. Although this may well be true, it doesn't help that public messages about recycling are always focussed on the same bit of the resource loop, the event of waste disposal. Some of the recent advertising has drawn attention to the entire journey of resources around the loop but the take home message is the same: don't chuck everything away.

This is a shame because the most interesting side of the recycling loop is not waste recovery but the design, manufacture and sale of products made from recycled materials. Although considerable government effort is going into developing markets for recovered waste, there is no campaign to enable or encourage consumers to buy reclaimed or recycled products.

We did our bit for the demand side of the loop this week when the walls of

*Sandis feeds the recycled newspaper blocks into the hopper.*

Tree House were filled with Warmcel 500, the insulation made from recycled books and newspapers. It was blown in damp then left to expand as it dried, ensuring that there are no gaps in the walls through which our precious heat might escape. We can also boast recycled aggregate and steel in the foundations, reclaimed floors and sanitary ware, roof tiles made from recycled plastic, and a garden fence made from scrap metal.

Perhaps the single biggest recycling resource in the UK is eBay, through which all the furniture for Tree House is being progressively sourced (our small Brixton flat has become a vintage furniture climbing frame, much to our cats' delight). eBay users, antiques dealers and car boot sale enthusiasts probably don't think of themselves as eco-warriors but the more we value the stuff we already have, the less we need to raid the Earth to make yet more.

All recycled materials have a latent mystery. When I am writing in the peaceful study at the top of Tree House, what stories of murder and mayhem will seep out of the walls and how soundly will we sleep within a cocoon of remaindered bodice-rippers? Reuse and reclamation make such questions even more explicit. Who relaxed on our

*The bedroom with cosy new walls.*

Chris Thompson and
his cabinet made from
reclaimed timber.

Aldis directs the
pumped insulation
into the walls.

sofa, laboured on our floors or soaked in our bath in the hundred years before they became part of our lives?

Such questions are of great interest to Staffordshire artist Christopher Thompson. Using reclaimed timbers from old potteries near his home, Chris recently sculpted a triptych of solid wood cabinets, each carved with a monumental front, including as a horse-head from the Parthenon frieze. Within each cabinet, there are memories of the workers and their craft: a teacup, sugar-bowl and cake plate made from the best Staffordshire porcelain. The cabinets express the strength and endurance of the materials; their contents, the fragility and preciousness of the lives that have encountered them. They will be the perfect adornment for our Warmcel-stuffed walls.

To put new heart into the green ring of recycling, we need to imagine the never-ending stories of all the things we fill our lives with. Including, of course, ourselves. I plan to be buried in a woodland and so become compost for a *Davidia involucrata* planted at my head. This tree, commonly known as the Handkerchief tree due to its papery blooms, will supply the mourners who gather beneath it with the perfect recycled product to meet their tearful needs.

*design detail*  **Kitchen bins**

To complement our use of recycled products, we do what we can to ensure that the resources that enter the house maintain their value after their primary purpose is complete. Significant space has been given over to recycle bins within the kitchen. We are keen cooks, so a single bin with small compartments for different materials would drive us crazy. Given the amount of space traditionally allocated to the storage of resources that come into a kitchen, it seems appropriate to reallocate some of this space to the resources leaving the kitchen.

These bins are for organic scraps, tins, plastic bottles, other recyclable plastics and the rest (landfill). Paper and glass are stored under the stairs.

*resources*

The monumental reclaimed cabinets were made by Christopher Thompson (01538 387001, chris@regentst.fsworld.co.uk).

The reclaimed teak floors and sanitary ware were supplied by Lassco (020 7394 2101, www.lassco.co.uk).

The Warmcel 500 recycled newspaper insulation for Tree House was supplied by Excel Industries (01685 845200, www.excelfibre.com).

The government programme to encourage markets for recycled products is called the Waste and Resources Action Programme (0808 100 2040, www.wrap.org.uk).

Useful online resources include www.recycledproducts.org.uk, www.reuze.co.uk, www.salvo.co.uk, www.recycle.mcmail.com, www.recycle-more.co.uk and www.greenspec.co.uk.

### Publication
*Cradle to Cradle: remaking the way we make things.* (W McDonough and M Braungart, North Point Press, New York 2002). A radical and inspiring vision of a world in which no resource ever becomes waste.

# *28 SEP 05*

For a year, communication on site had never been much of a problem as the few pieces of plywood that had sufficed for our makeshift floorboards left plenty of gaping holes. The installation of the upper floors, albeit without their finishes, took us a significant step nearer to a building that felt like a house that might one day be a home.

The floorboards came with pre-installed underfloor heating pipes within them, a clever system that is ideal for new build. On the solid ground floor, a single pipe was laid on special insulation panels, then a screed poured on top. Although the pipes were pressure-tested for leaks, they could not be tested with warm water until the heating system was commissioned. Unfortunately, this important event was still months away.

## Underfloor heating

Are you a hothead with a predilection for getting cold feet? If so, it's time you did something about your personal temperature gradient. This is not some faddish self-help wheeze but the tricky challenge of making the heat in our homes match our delicate human

*Genc and Nik lay the floorboards with pre-installed heating pipes, taking great care where their screws go.*

sensibilities. Whatever interior climate you like – be it the soul-enhancing Scottish croft experience or the always-on sauna style – you will feel more comfortable if your feet are warmer than your head. If you can toast your toes at the fireplace, your brain will glow contentedly even in quite a cool room.

Despite this basic physiology, most houses in Britain employ wall-mounted radiators to supply their central heating, a system that works precisely contrary to this ideal temperature gradient. Radiators deliver most of their heat not through radiation but through convection, creating upward air flows that drive warm air to the top of every room and suck in cold draughts at floor level. The result can be rooms that are stuffy, uncomfortable and prone to cold spots, rooms where 'putting your feet up' may be the only way of getting thermally comfortable.

This clumsy approach to heating is far too inefficient for 'zero carbon' Tree House. Instead we are using a system that will deliver room temperatures that almost exactly match our bodily preferences: the admirable Roman invention of underfloor heating. Although the seductive idea of the warm floor has changed little, the technology has moved on. Roman hypocausts used warm air blown into a subfloor from an adjacent furnace, whereas modern systems rely on piped hot water or, for a lot more carbon emissions, electricity.

Unlike the misnamed radiators, a warm floor heats a room principally by radiation, so you get a very consistent, stable output with no swirling currents of air. Because the heat is greatest where you appreciate it most, you don't need as much heat to reach the same level of comfort, and because the water in the pipes does not have to be as hot as in radiators (50°C rather than 80°C), your boiler can operate at a more efficient temperature. This is especially important for condensing boilers that

*The pipes are all connected along a central channel in the floorboards.*

work best when the water returning to them is cool. It is ideal for our ground source heat pump, which cannot supply the scalding temperatures demanded by radiators.

A big advantage of the system we are installing is the integration of the pipework and flooring in one product. This made it possible for Genc and Nik to do two jobs at once: installing the underfloor heating as they laid the floorboards. On our solid ground floor, we will be threading the pipework on to special insulation panels before laying a concrete screed on top. When complete, the other great benefit of underfloor heating will become apparent: it does the job without getting tangled up in your furniture.

Because the system works by heating up a radiant floor, underfloor heating takes time to respond to changing needs. It therefore works best in frequently occupied, well-insulated houses that don't require a lot of heat quickly to bring them up to comfort temperature. Not worth the money if you're never at home.

We will be finishing our floors with reclaimed teak upstairs and Kirkstone slate on the ground floor. The result should be stunning, even if it doesn't quite match the extravagance of the Romans whose warm floors were lined with mosaics. These were doubtless admired by gentlemen farmers and soldiers alike as they snacked on imported delicacies or applied the latest oils to their weary limbs.

*On the ground floor, one continuous pipe is laid in special insulation panels before a screed is laid on top.*

173

What did the Romans ever do for us? Interior design, the Mediterranean diet, moisturising cosmetics and the perfect personal temperature gradient. Two millennia later, it seems our faddish modern lifestyles are still only just catching up.

## Room thermostats

Room thermostats ensure that one end of your home does not get too hot when the other is still warming up. In a very low-energy house with lots of solar gains, this is especially important – the east-facing room at the top of Tree House is quickly heated by the morning sunshine while the west facing ground floor living room is still benefiting from the underfloor heating.

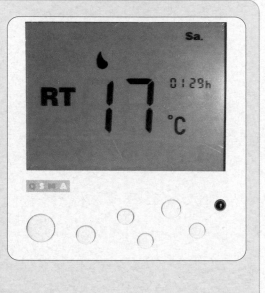

The underfloor heating for Tree House was supplied by OSMA (01249 766 600, www.osma.co.uk).

### Publications

*Environmental Science in Building* (R McMullan, Palgrave 2002). If you want to gain a more thorough understanding of how heat, air, light, sound and water flow through buildings, try this accessible textbook.

*A Guide to the Roman Remains in Britain* (RJA Wilson, Constable 2002). Locates and describes every extant Roman remain in the country, including all those hypocausts.

# 05 OCT 05

The installation of the photovoltaic modules on the roof of Tree House was a turning point in the build. Until this moment, the project had done nothing but consume energy and resources. Now, for the first time, we were producing energy and giving something back. Although the internal electrics in the house had not yet been fully wired, our roof-top power station was fully Grid-connected and exporting power from day one.

The design, installation and commissioning of our power station was undertaken by Solar Century, working with our architect, Peter Smithdale. They coped admirably with our ever-changing schedule and did a thorough job on-site. It is a really great pleasure to live beneath their sparkling solar roof.

*Special plastic panels are fitted on the roof to act as a weather shield and the foundation for the PV modules.*

# Photovoltaics

At the front of our little urban building site, the hoary old sycamore that dominates our plot has had a good summer despite all the aggravation beneath it. Up in the crown of the tree, each fragile leaf spreads wide its five fingers under the October sunshine, creating sugars for new growth. This process is quietly miraculous, for the entire towering structure of the tree is self-built using these flimsy solar collectors.

To my delight, the canopy of Tree House has finally caught up with the tree that it seeks to emulate. All last week a team from Solar Century was busy installing thirty square metres of photovoltaic (PV) modules on our roof. On Friday, Trevor and Guy connected our inverters to turn the modules' DC output to 240V AC and the final switch was thrown. It's taken a long time to get here but the moment is sweet: from this point on, we plan to consume no more power within our narrow Clapham boundaries than our glittering solar roof can supply.

At the beginning of the project, the goal of energy self-sufficiency seemed beyond our reach. The site was small and constrained and many PV suppliers suggested that a rooftop array could typically cover no more than half of annual power demand. But after some serious research, reflection and arithmetic, the possibilities of an atypical home became clear: an ultra-efficient house with a big southern-pitched solar roof could go all the way, supplying not only our lights and appliances but also the power needed to drive our solar thermal panel and heat pump (for hot water and central heating). The goal is tough, but we are one big step nearer to achieving it.

PV is an extraordinarily under-developed resource in the UK. This is odd given that PV modules are silent, emission-free, unobtrusive, low maintenance, reliable and easy to integrate into buildings of all sizes. The main problem has always been cost: no one currently installs a PV array on their roof on the basis of a payback calculation. The cost of PV has fallen but we could not have

*The modules are slotted into place.*

fulfilled our ambition without a 50% government grant. We are also fortunate to have financial support from Halifax and Bank of Scotland, who are keen to promote good environmental design in the development of homes old and new.

Arguably PV is a technology still lingering in the 'Valley of Death', the rocky terrain between technology development and large-scale commercial success that is notoriously hard to cross. Elsewhere – Germany, Japan, the Netherlands – governments have been quicker to propel PV towards the verdant uplands, often by paying domestic PV generators artificially high rates for the power they produce. In Britain we won't be making a killing from our roof-top power station, though having zero energy bills at the start of a long-term rise in energy prices is no bad thing.

There's little point in investing in PV if you haven't done all the other energy-efficiency jobs first. But if, like us, you want to travel somewhere beyond the typical, PV is the renewable power source with the most universal application.

Unlike David Hume, I accept that the sun will not only rise but set every day, and hence that PV can only be part of the answer – we will be grid-connected, exporting power in the day and importing at night. But as the answer to our planet's fossil fuel dependency is needed with some urgency, every little fragment of renewable energy capacity is invaluable. Today, Tree House is exceptional. One day, it will be an old oak amid a city-wide forest of environmentally benign solar buildings. I hope that we are around to see the day: a couple of hoary old woodsmen savouring the moment with long-awaited satisfaction.

---

*design detail* | Garden LEDs

The panels on our roof use silicon-based semiconductors to generate electricity. In our small courtyard garden, a very different semiconductor design expends energy: the light-emitting diodes (LEDs) of our garden lighting.

As our ground floor living space is very integrated with the garden, we did not want to lose half the space whenever night fell. Nor, however, did we want to use a lot of energy lighting a space that we would often only look at. The solution is to use efficient, low-powered technology very carefully. Remarkably, each of our seven LED micro-spotlights only consumes one watt of electricity.

## Al Green

Al was Solar Century's project manager for Tree House, overseeing the design, specification, installation and commissioning of our 5kW photovoltaic roof. He had to deal with more than his fair share of rethinks and postponements on this project but remained supportive, helpful and patient throughout. Very much appreciated.

The photovoltaic roof for Tree House was designed, supplied and installed by Solar Century (020 7803 0100, www.solarcentury.co.uk).

For information on grants for domestic photovoltaic installations, contact the Energy Saving Trust (www.est.org.uk, 0800 915 7722).

The British Photovoltaic Association is a trade organisation providing news and information about the technology (www.pv-uk.org.uk).

### Design detail
Auraleds supplied by Light Projects (020 7231 8282, www.lightprojects.co.uk).

### Publications
*Ecohouse 2* (S Roaf, M Fuentes and S Thomas, Architectural Press 2003). Includes a detailed chapter on photovoltaics.

*Practical Photovoltaics* (R. Komp, Chelsea Green 2001, available from Green Books). Classic reference on solar electricity.

# 12 OCT 05

By mid October the space that would become our laundry was clear enough but the room was still entirely imaginary. The concrete screed went down but was walked on too early and had to be resurfaced. The walls and ceilings were full of insulation, but not yet covered in plasterboard. The window was in place, but the door was just a hole awaiting its frame.

Nonetheless services were going in and second fix items (such as taps) were being contemplated, if not actually fitted. With a complete structure to work with, the project was advancing on a number of fronts at once. Every day there was something new to appreciate.

## The laundry

Design can save the world. Or rather, design could save the world if we created the right incentives for sustainable design to flourish. We could begin by taxing traditional lightbulbs and their inglorious 2% efficiency right out of the market-place. But even then, can we really continue our full-on lives without turning anything off? Do we need to redesign our lifestyles as well as our buildings and technology?

*More insulation is fitted, this time on the underside of the roof.*

I am convinced of the potential of design to make a huge difference to our hidden ecological impacts, but I also recognise that sustainable design often complements changes in behaviour and lifestyle. This isn't an either/or but a subtle interplay in which design and behaviour work together to create new worlds.

I've been contemplating these new worlds at a humble level in the specification of the small laundry for Tree House, which will soon fill out from its newly laid screed floor. Is it enough to buy the most efficient washing machine and tumble dryer and just carry on as usual? Well, no. Thorough-going eco-design requires that we consider the whole process, rather than just tweaking the details.

Laundries have been banished from modern homes by the ascendancy of white goods, often turning washing into a mechanistic chore rather than a task undertaken with thought and care. The design decision to include a dedicated laundry space in Tree House is therefore a deliberate attempt to improve both our technology and our actions. On one side of the room will sit our washing machine, on the other a large reclaimed Butler sink. How many of the items in our laundry basket will really require a machine wash and how many will be happy with a quick cool stir? After all, we're a couple of urban professionals (albeit male ones), not coal miners.

Although we may not always choose to use it, our AEG-Electrolux washing machine is itself an eco-design success story. It requires only 39 litres of water per wash, a third of the consumption of a ten-year-old machine, and only one kilowatt-hour of electricity. It is also designed to optimise our behaviour by incorporating a weight sensor that indicates when a full load is reached. This is important because most people consistently underload their washing machines and so end up using them more often than they need to.

Then we face the tricky issue of drying. Here there is no eco-design quick fix: even an A-rated tumble dryer will gobble lots of energy to perform a task that takes place naturally for free. Gas-fired dryers have much lower carbon emissions than the familiar electric models, but they still can't match the environmental credentials of the humble washing line.

Although our small back garden will have a retractable washing line, even this simple technology fails in the winter. So above our washing machine we will be installing a brilliant piece of Victorian eco-design, the clothes airer. This simple contraption is lowered to hang damp clothes then raised on pulleys to make the best use of the warm air collecting at ceiling height. As few high

*A Victorian clothes airer has pride of place in the Tree House laundry.*

street shops sell them, I purchased our Victorian original through eBay just before it got thrown into a skip. You can, however, buy a contemporary clothes airer and plenty of accompanying eco-friendly detergents from the Natural Collection.

Hopefully, the laundry of Tree House will turn good design into good behaviour: within a few weeks, Ford and I expect to be (need you ask?) the best-behaved laundry boys in south London.

## design detail

### Peace lily

Plants are remarkably good at cleaning the air, absorbing toxins through their leaves and breaking them down in their roots. This is an important function of trees in urban environments and of houseplants within our homes.

In the 1980s research by NASA identified a range of plants that were particularly good at cleaning the air. *Spathiphyllum*, the peace lily, absorbs formaldehyde, benzene and carbon monoxide. Although pot plants will never replace careful ventilation as a primary means of maintaining air quality, this air-cleaning role beautifully complements their other aesthetic, calming and humidifying functions.

## resources

The washing machine for Tree House was supplied by AEG-Electrolux (08705 350 350, www.aeg-electrolux.co.uk).

Clothes airers and many other ecoproducts are available from the Natural Collection (www.naturalcollection.com, 0870 331 33 33).

### Publication
*Natural Stain Remover: Clean your home without harmful chemicals* (A Martin, Apple Press 2003). Concise advice on traditional non-toxic cleaning methods.

# 19 OCT 05

The little courtyard garden I planted at the back of Tree House gave a final late summer flourish before subsiding, long before we were able to admire it from our sofa as planned. Never mind: gardening is all about patience and I was happy to rethink, replant and look forward to the next year.

Our front garden never had a chance of life during the build, except for our all-embracing tree, which thrived despite all the stomping, digging and dumping that went on at its base. I knew that this space would take much longer to turn into a mature garden than our little courtyard at the back, not least because of the dry shade conditions under the tree canopy, but I was determined that it should feel just as special as the rest of the house. The focus on wildlife gave us a clear direction for its design. The front garden was actually occupied at this time by Hughie the metal-worker, cutting and folding our elaborate roof. Inside the building, the long slog of plasterboarding had begun.

*In the space that will become our wildlife garden, Hughie fabricates the stainless steel details of the roof.*

*The interior plasterboarding begins.*

# Wildlife gardening

I'm standing on the top-floor balcony of Tree House enjoying the late autumn light in the crown of our big old sycamore. Finches and sparrows are hopping between the bird feeders in the lower branches and the protective cover of the bushes below. The English hedgerow at the front of the garden is full of the fruits of autumn and I know that a multitude of insects are clambering around the log pile at the base of the tree and through the ivy climbing the walls of the house.

The only thing that spoils the picture is the inescapable fact that I've made it up. Nonetheless, like all my fantasies of life in Tree House, this vision is beginning to rise above my dream horizon and brighten our quiet corner of Clapham for real.

The end of the build is almost in sight. Last week, we gave our notice to the landlord of our Brixton flat, so come hell or high water (in our case probably both) we will be moving in at the end of November. It will not be finished, but if we have a water supply, toilet and staircase, a basic standard of living should be possible. We might even survive without a staircase, though this would be a challenge for our cats. Ford has trained Trevor to stand on his hind legs and beg for food but it would take more than roast chicken to get him up a ladder – Trevor that is, not Ford.

With all efforts focussed on completing the building, I am wary of losing sight of the vision for the gardens that are integral to the design. So last week, I paid a visit to the Centre for Wildlife Gardening, run by the London Wildlife Trust, to get some local advice about attracting birds, insects and other wildlife to our front garden (frogs to the pond at the back please).

*First planting – cotoneaster will provide berries for the birds through the winter.*

Moya O'Hara showed me around, pointing out the many easy ways of making urban spaces wildlife-friendly. First, plant a native hedgerow (predominantly hawthorn with lots of interesting extras) as this is a rich source of nectar, fruit and seeds. Thorny shrubs also provide good cover for birds when feeding, so it's worth siting manmade birdfeeders within their woody framework, though not completely out of sight. Non-native berberis, firethorn and cotoneaster are also good for berries and protection.

Second, dead wood provides an excellent habitat for many creatures. We are thinking of making an entire garden fence out of logs, though we will also want some half-buried logs as these are particularly attractive to stag beetles.

Third, grow herbaceous plants appropriate to your soil and shade conditions that will provide nectar, seeds and berries across the year. Avoid cutting them back in the autumn as the dead seed heads are still a source of food and a habitat in themselves. The woodland planting at the base of our tree will include native English bluebells in the spring followed by foxgloves, cranesbills, bugle and red campion.

Fourth, plant native ivy, an exceptional wildlife plant that provides nectar and berries and is a great habitat for birds and insects alike. It will perfectly complement the sharp white cladding of Tree House.

Fifth, plant a mini-meadow of long grass and wild flowers where grasshoppers and crickets can live and butterflies can lay their eggs. This is not easy in a shady space, but we might just squeeze one in if there is a corner that catches the sun.

I left Moya and the birdsong of Peckham to return to the harsh reality of our building site. It may be years before the view from the balcony is fully fledged, but it will be worth the wait. Although our cats will not have access to the front garden, they will have their own place in this furry, as well as feathery, vision: lined up beside me on the balcony, grinding their teeth.

*Moya O'Hara with woodpile and native hedge in the Centre for Wildlife Gardening.*

## Marble birdbath

This must be the heaviest birdbath in London: a great chunk of marble with a simple concave cutaway where the birds visiting our front garden can preen.

The birdbath is positioned in line with the laundry window. As our salvaged Butler sink is positioned on the other side of the window, we can wash our socks while watching the birds wash their feathers.

The birdbath is also salvaged, picked up at auction in north London for considerably more than I wanted to pay. Happily the hole in my wallet is a distant memory, whereas this birdbath will last forever.

The London Wildlife Trust (020 7261 0447, www.wildlondon.org.uk) has produced an excellent free pack on wildlife gardening and provides details of nature reserves across the capital.

### Publications

*Attracting Birds to Your Garden* (S Moss and D Cottridge, New Holland 2000). Well illustrated with a species gazetteer.

*How to Make a Wildlife Garden* (C Baines, Frances Lincoln 2000). The established guide.

*Wildlife Friendly Plants: Make your garden a haven for beneficial insects, amphibians and birds* (R Creeser, Collins & Brown 2004). Detailed information on plant selection for wildlife.

# 26 OCT 05

At the end of October the regular team was fully occupied plasterboarding the walls and ceilings. This took time, partly because we had quite a few tricky corners and curves to deal with, especially at the top of the house, but also because everything had to have a double layer of board. The two layers were designed to give us improved fire protection and to add 'thermal mass' to our lightweight house, helping to reduce the risk of summer overheating.

There were also frequent visits from other trades including plasterers, electricians and plumbers. But despite all this activity, I was becoming increasingly worried that our latest completion date – the end of November – was not to be. The lack of a staircase, or any evidence of a staircase, was the principal hole in my confidence.

*Nik works on the plasterboard details around the VELUX windows.*

*The plumbing is carefully fitted through the joists.*

# Indirect energy consumption

God, I'm exhausted. We're three years into our ambitious, extravagant self-build; we've been on-site for thirteen months; the costs have gone through our glittering solar roof; and we're supposed to be moving in at the end of November despite the fact that the first fix plumbing has only just begun.

I always have this problem at 95%. The light at the end of the proverbial tunnel drains my energy and I cast about in the gloom for an escalator to take me up the final 5%, only to find the staircase getting even steeper. If ever I needed to improve my personal energy efficiency, it's now.

Saving energy is not just about insulation and low-energy lightbulbs. Beyond our immediate needs for heat, light and power, there are many indirect ways in which we can still guzzle energy. Under our beautiful super-insulated roof, currently being completed by millimetre-accurate Nik, we will also be saving indirect energy wherever we can.

For starters, lots of energy is consumed purifying and pumping mains water, so we are keeping our demand low by installing taps and showers with aerators that provide a fuller flow for less water, ultra low-flush Ifö ES4 toilets, water-efficient appliances and an underground rainwater tank to supply our garden and pond.

Building materials also come with substantial 'embodied energy' – the energy expended in extraction, manufacturing and transport. Consequently,

*Will's allotment keeps his food miles down and his quality of life up.*

although our foundations are full of carbon-intensive concrete, above ground we have prioritised natural, unfired materials such as wood and English stone.

Whenever you buy new stuff of any kind, these hidden costs clock up. To avoid them altogether, buy second-hand. We're using a wide range of reclaimed materials in the build, all our furniture is being sourced through second-hand shops and eBay, there's a great second-hand bookshop in Brixton (Bookmongers) and I often buy good quality clothes from TRAID (Textile Recycling for Aid).

Even more problematic is the national and international transport of food, given the volume we all consume. We plan to keep our food miles down in Tree House by tending our allotment, growing herbs in the back garden, boycotting supermarkets' centralised distribution networks and buying local produce whenever possible, especially at farmers' markets.

Finally, waste is a waste of energy because whenever you send something to landfill you are also throwing away all the energy that went into making it. We will be recycling almost everything with a wormery in our back garden and multiple recycle bins in our kitchen. We also plan to reduce our packaging waste by combining bulk-buying with a glamorous display of large Bodum storage jars.

Is that enough energy saved to get me through the last 5%? One last burst of solar-powered enthusiasm and I'll be sitting back, admiring the view and wondering what I can possibly build next.

*The roof, nearly complete, with photovoltaic modules and solar thermal panel.*

*design detail*

## Aerating tap

Any tap with a screw thread at its outlet can be fitted with an aerator. This gives a fuller flow for less water and therefore less energy.

*resources*

The water-efficient taps for Tree House were supplied by Hansgrohe (0870 7701972, www.hansgrohe.co.uk).

The Ifö ES4 low-flush toilets were supplied by the Green Building Store (01484 854898, www.greenbuildingstore.co.uk)

The Energy Saving Trust is the first port of call for advice about energy saving in your home and car (0800 915 7722, www.est.org.uk).

The Food Climate Research Network collates information on the hidden energy impacts of food consumption in Britain (www.fcrn.org.uk).

To find a farmers' market near you, contact the National Association of Farmers' Markets (0845 230 2150, www.farmersmarkets.net).

For ideas on recycling a wide range of domestic goods see www.reuze.co.uk. Local recycling banks can be located through www.recycle-more.co.uk.

### Design detail
Aerating tap by Hansgrohe (www.hansgrohe.co.uk)

### Publications
Conserving Water in Buildings. Eleven concise fact sheets on water efficiency in homes, available from the Environment Agency (08708 506 506, www.environment-agency.gov.uk).

Wise Moves: exploring the relationship between food, transport and $CO_2$ (T Garnett, Transport 2000, 2003). An analysis of the problem of food miles in Britain.

## 02 NOV 05

I was determined that the building should be pressure-tested, given our commitment to significantly cutting our energy losses from unwanted air leaks. The tricky bit was the scheduling: there is little point in pressure-testing a building which has windows or roofs missing but, if you leave it too late, the holes in the fabric become impossible to find or treat.

As it happened, there were still two gaping holes when we ran the test, the glass block window in the bedroom and the small glazed roof in the corner of our main living space. We covered and taped these up as best as we could and when the moment came to turn on the fan our very temporary seals held – just.

# Pressure-testing

Paul Jennings is not your everyday man with a van. He is, in fact, a man with a fan in a van. What's more he is a man with a plan for his fan in a van. In my humble opinion, albeit lacking the imprimatur of *The Cat in the Hat*, this plan of the man with the fan in his van has a certain – élan. And these are the reasons why:

*Paul Jennings checks the fan before the pressure test.*

*Paul checks for leaks around all the windows . . .*

*. . . and where services penetrate the building.*

Paul works for a company called Stroma Technology that pressure-tests buildings of all sizes to see how airtight they are. An airtight building can be carefully ventilated to minimise heat losses and optimise indoor air quality. By contrast, a building that is not airtight will leak warm air through cracks and gaps, at worst leading to draughty, cold and uncomfortable interiors.

Builders in Britain have never paid much attention to airtightness. Today new houses have to be built with much more insulation than in the past but they can still be full of holes. Although there has been a standard of airtightness in the building regulations since 2002, there has been no requirement for new houses to be tested to see if they achieve it. This will finally change in the 2006 revision to the Building Regulations.

Tree House, of course, has got to be different. Our ever-patient architect, Peter Smithdale, tracked down the world's best airtightness detailing for timber-frame houses (from Canada) and did his best to design them in. Unfortunately, it's not easy to build an airtight house because the downright fiddly business of sealing every junction, crack and service duct is ill-suited to the hurly-burly of a building site. Nonetheless, our guys on site have been attentive, so we were hoping for a good result.

We closed all the windows and external doors and Paul sealed the front door with special panels into which the fan was inserted. Then the fun started. With the fan on and the house under increasing pressure, we were able to go round the building with a little smoke puffer looking for action. Junctions where the smoke made little quiet clouds were good news; places where the smoke whipped away out of sight were trouble. Happily we can still take a lot of remedial action to deal with the hot (or rather cold) spots before the build is complete.

Then came the crucial measurement. The fan was ratcheted up until the pressure inside the house was a steady 50 Pascals (Pa). At this pressure, Paul measured the air flow out of the house and used this to calculate the 'air permeability' of the building: the volume of air that gets flushed through the external envelope of the building every hour. The standard in the building regulations is a maximum of 10 m3/m2/hour at 50Pa. We achieved 3.4, and with remedial action may get this down quite a bit lower. Not bad. In fact, really quite good. Especially for a tricksy design with lots of fiddly corners.

As this level of airtightness would not give us enough fresh air at normal air pressure, in the winter we will be mechanically ventilating the house using a heat-recovery unit that changes the air but keeps the heat in. In summer, we will open the windows. Paul reckons that Tree House is likely to be better than the current output of the volume housebuilders by a factor of four or five.

If you want to plan for Paul's fan in a van, think about airtightness from the very start of your project. Your reward will be a very comfortable draught-free house at the end of it.

## Cat doors

As well as letting cats in, a typical cat door also brings huge quantities of cold air into a home. An airtight house can be completely ruined by its furry inhabitants.

We will have two cat doors. Ordinary cat doors do not have a good seal as they let cats both in and out of the building. Instead of this flapping at all hours arrangement, we will have an 'in' door and an 'out' door, each of which will have a proper seal, the door closing tight against a nib.

The doors will open onto little feline stepping-stones across our pond. Hours of entertainment guaranteed.

Stroma Technology pressure tests buildings of all sizes from houses to aircraft hangars (01924 870 677, www.stroma-ats.co.uk).

The LoWatt heat-recovery ventilation unit for Tree House was supplied by Vent-Axia (01293 526062, www.vent-axia.co.uk).

### Publications
The following Best Practice in Housing publications are free to download or order from the Energy Saving Trust (0800 915 7722, www.est.org.uk):

*Energy efficient ventilation in housing* (GPG268)

*Improving air-tightness in dwellings* (CE137/GPG224)

*Post-construction testing – a professional's guide to testing housing for energy efficiency* (CE128/GIR64). Introduces pressure-testing as one of many tests to check building performance.

# 09 NOV 05

After another difficult meeting with our contractor, I had to return to our landlord, cap in hand, to ask if we could stay in our Brixton flat for one more month. Fortunately this was possible, so we did not have to make plans to put our belongings – and cats – in storage. Nonetheless it was one more expensive month to fund at a time when we were already reeling from unexpectedly high bills for the completion of Tree House.

On site, the installation of the base units of our kitchen gave us our first real sense of how well the open-plan living space on the ground floor would work. Although we had deliberately given a lot of space over to the kitchen, the use of floor-to-ceiling glazing onto our courtyard garden meant that the space still felt remarkably big despite everything we were trying to squeeze into it.

*Progress on the plumbing front.*

*Stephen Edwards installs the base units of the galley kitchen.*

# Cold appliances

Mohammed Bah Abba is something of a local hero in the arid plains of northern Nigeria. He is credited with transforming the lives of subsistence farmers, invigorating the local pottery industry and releasing women from the burden of daily treks to the vegetable market. Mohammed is not, however, a radical social reformer but an inventor, and the invention which has cut the local mustard so convincingly is his 'pot-in-pot' or the powerless fridge.

*The energy-efficient, sensibly-sized fridge in the Tree House kitchen.*

The pot-in-pot works by evaporative cooling. One large clay pot is placed inside another with a boundary layer of sand between the two. Tender crops are packed into the inner pot and covered, then water is poured into the sand. The water slowly evaporates, drawing the heat from the contents of the pot and keeping the vegetables fresh for days.

The same principle is used in simple cooling devices the world over. A company in Ohio has been making a passive water cooler for over a century, based on an Amish design. This is simply a porous clay urn that quietly sweats its contents, helping to keep what remains cool. Terracotta wine coolers work in the same way, as long as you remember to give them a good soaking first.

There is a limit to the potential of such simple evaporative coolers in the kitchens of Britain. After all, a pot-in-pot is never going to look as good as a six foot high bright pink retro-chic fridge-freezer, the unassailable accessory of all domestic food designers worth their coarsely ground sea salt.

Perhaps, though, there is a middle ground. This is what we are seeking at Tree House where the outline of our kitchen is taking shape amid the general hubbub of wiring, plumbing and plasterboarding. With the FSC birch-ply carcasses installed, Ford and I have had our first impression of life in the culinary laboratory that looks over our living space, through the glazed boundary of the house, across the pond and into our courtyard garden. If this view should ever distract Ford from the razor-sharp preparation of our favourite Japanese raw fish dinner, we will be able to pop his amputated fingers into our energy-efficient under-counter fridge while we wait for help to arrive.

The most important eco-choice when buying a fridge or freezer is to check the energy rating: nothing less than A is acceptable and A+ or A++ is desirable (the energy-saving logo is now only given to appliances rated A+ or better). Regrettably, however, the energy rating can be misleading because it is based on the energy use per cubic metre of space inside. This approach favours bigger appliances because as the size of a fridge increases, the additional energy required to cool the interior decreases. But extra energy is still extra energy, so an A-rated mega-fridge will use far more power than a B-rated under-counter fridge.

*The view from the chopping-board – across the pond into the garden.*

So don't just check the energy rating when making comparisons between products, check the annual energy consumption too. Then think very hard about how much powered cooling you really need. Really really.

If any entrepreneurs out there think that the pot-in-pot might catch on here, do get in touch. A little retro-chic detailing and I'm convinced we'll be millionaires.

## design detail

### No 137 bus

The number 137 bus stops just around the corner from Tree House. On a Saturday morning it takes us direct to Pimlico Road farmers' market where we can supplement the fruit and vegetables from our allotment with high-quality produce that has not been shipped around the world or around the country. Given the extraordinary distances that food is transported in our globalised world, this will save far more energy and carbon emissions in the long term than our choices of efficient fridge and freezer.

## resources

The energy-efficient fridge and freezer for Tree House were supplied by AEG-Electrolux (08705 350 350, www.aeg-electrolux.co.uk).

The Energy Saving Trust maintains a database of energy-efficient appliances. These can be identified in shops by the 'Energy Saving Recommended' logo (0845 727 7200, www.est.org.uk). Only A+ and A++ fridges and freezers are given this recommendation.

For more details of the pot-in-pot and other innovative ideas for a better world, see www.worldchanging.com.

# 16 NOV 05

Although the interior of Tree House still looked a complete mess at this belated point in the project, the exterior was really beginning to come together. The roof was all but complete and looked stunning, the gleaming photovoltaic array to the south complemented by the muscular roof windows to the north. At the front of the building, the red cedar cladding for the stair tower was going on, creating the long-awaited trunk of Tree House. At the back, the small western roof over the extended living room was looking attractive finished in cedar shingles.

This diary entry was the culmination of a long design process. We were committed to using bicycles and getting rid of our car but the very tight constraints on our building footprint – we could not build within five metres of the tree – and the exposure of the front garden to the street meant that we had struggled to find a secure way of storing bikes that did not spoil the immediate experience of the interior of the house. We eventually realised that the solution lay not with the design of the building but with the design of the bikes.

*The northern roof windows look stunning next to*
*stainless steel flashings and the recycled slates.*

*The cedar cladding for the trunk of Tree House
begins to wrap round the stair tower.*

*Cedar shingles on the small western roof.*

# Transport

As Tree House nears completion I am optimistic that the outcome of our long months of thought, design and labour will be something more than an interesting, sustainable building. The house will, I think, be beautiful.

I would like to argue that sustainability and beauty are ineluctably linked, but of course they are not. There are plenty of ugly eco-houses in the world and beauty is resolutely value-free: environmental degradation can be beautiful too, as the film 'Koyaanisqatsi' illustrated brilliantly for 1980s America. Nonetheless there is a particular beauty inherent in sustainable design that, in the right hands, shines through.

As we want Tree House to capture this beauty in every detail, one of our largest possessions will have to be sold before we move in. On any aesthetic reckoning, a battered K-reg Nissan Primera is going to be too darn ugly for the front garden of Tree House, to say nothing of the obvious impact it will have on our low-carbon ambitions.

It's taken time to work out the alternative but we are confident that we will not be giving up anything major beyond the expensive beast itself. For starters, the house is in walking distance of three tube stations, two dozen bus routes, Brixton market, Clapham High Street and our allotment. Cities like London often get criticised for their environmental impacts but this kind of localism is often much harder to achieve in the sprawling 4x4-filled countryside.

*Terrence assembles a folding bicycle in Brompton's west London factory.*

It is, however, difficult to replace the flexibility of a car, not least for getting out of the city every now and then. Our solution lies in a factory in west London where the only vehicle still built in the capital comes off its tightly run production line: the Brompton folding bicycle.

The Brompton is a design classic, a brilliant piece of engineering that rapidly folds up into a neat, lightweight bundle that can go with you anywhere. It solves the flexibility problem because we can get on any train whenever we want and still have transport at our destination. This is actually more attractive, at least in good weather, than trying to escape south London by car. It also solves the problems of bicycle security and storage. Ideally every new house should be built with a secure porch, outside the insulation envelope, where bikes (and vegetables) can be stored, but all too often there isn't space for this in tight urban developments, including our own. With folding bikes, all we will need is a cupboard under the stairs.

Business is booming at Brompton where Terrence and team assemble over seventy bicycles every day. Hopefully this reflects a growing awareness that there are more ways of getting from A to B than a choice of drivers' rat runs. If, however, you don't want to completely spurn life behind the wheel, join a car club for the occasional nostalgic trip.

Personally, I think the Brompton is a beautiful product. It looks good and it does good. It is deservedly one of the exemplars in 'The Total Beauty of Sustainable Products' by Edwin Datschefski, a book that makes a persuasive case for the convergence of beauty and sustainability.

If we succeed in capturing this convergence in Tree House, it will in large part be due to our starting point: the tree. Although strictly it has no designer (sorry Mr President), is there any better example in our everyday lives of the total beauty of sustainable design?

## *design detail*

### Ligne Roset Togo sofa

This sofa did not make it into Edwin Datschefski's book. Nonetheless it is a) a design classic and b) second-hand, bought through eBay. So although it may not be an exemplar of sustainable design, it is still a beautiful, sustainable purchase. Almost all the furniture for Tree House was bought second-hand – always the most ecological choice.

The striking organic form of this sofa makes it the perfect nest for our feline dependents on the top floor of Tree House.

## resources

Brompton folding bicycles can be tried at dealers throughout the UK. For details, contact Brompton Bicycle Ltd (020 8232 8484, www.bromptonbicycle.co.uk).

*A to B magazine* focuses on folding and electric bicycles and alternative means of travel (01963 351649, www.atob.org.uk).

To find a car club near you, contact CarPlus (0113 234 9299, www.carplus.org.uk).

Transport 2000 is the national campaigning organisation for sustainable transport (www.transport2000.org.uk).

The Man in Seat 61 will tell you how to travel the world without flying (www.seat61.com).

### Publications

*City Limits*. A resource and ecological footprint analysis of Greater London (Best Foot Forward, September 2002). The last word on the ecological impacts of the capital.

*Cycling in the UK: The official guide to the National Cycle Network* (N Cotton and J Grimshaw, Sustrans 2005). How to make the most of the new national cycle network. See also www.sustrans.org.uk.

*The Total Beauty of Sustainable Products* (E Datschefski, Rotovision 2001). A clear model of sustainability is used to consider the virtues of a wide range of products. Edwin Datschefski also has his own website: www.biothinking.com.

# 23 NOV 05

The plan for the inside of the house was to complete from the top down, clearing the mess and applying the finishes as we went. Once the top room of the house had received its plaster skim, the big exposed roof trusses were sanded down ready for an oil finish. This brought out the colour and rich patterning of the wood and also provided protection for the next stage, the paint job.

Ford and I planned to save a little money from our burgeoning budget by applying the finishes ourselves. As we were using entirely natural products, we could enjoy a pleasant range of smells along the way. The oil for the woodwork had a citrus solvent, so the house smelt appropriately of oranges as Christmas approached.

| Calvin puts the finishing touches to the plaster skim in the bedroom. | Pete sands down the big exposed timbers in the study before they are sealed. | Richie shows off his technique. |

## Paints

There's nothing like a wall finish to bring joy to the heart of a self-builder. I've grown so accustomed to the guts of Tree House hanging out all over the place that a perfectly plastered wall, courtesy of Calvin and Richie, feels like an extravagant luxury. In fact, it's the cue for more hard labour – a long anticipated paint job.

Whitewash, a traditional paint made with lime, is the etymological ancestor of greenwash, the corporate practice of dressing up a polluting business in a cloak of environmental respectability. Although this is irresponsible, I think far greater harm is done by pinkwash: imbuing environmentally harmful products with such a reassuringly fluffy aura that the ordinary shopper has no hesitation in snapping them up and foisting them on their new-born babes.

*The study ceiling took forever to paint but the end result is light, bright and a daily inspiration.*

Paint is a pinkwashed product. Little Dorothea is due in two weeks, so there's just time to pop down to the DIY warehouse and wander among the bright rows of gleaming tins, browsing the colour catalogues filled with cheerful homes and healthy families. In no time mum and dad are back home with ten litres of Oxtongue Cerise to adorn the nursery walls. Dad gets a headache painting the room but the air soon clears and all that remains is the satisfying smell of freshly painted walls.

Such pinkwash is highly effective but you don't have to scratch much away to see what lies below. Synthetic paint is made from crude oil, cracked at high temperatures to produce a range of hydrocarbons that are then processed to make a fistful of chemical ingredients, leaving a volume of toxic waste behind many times greater than the paint produced. When the paint is applied, the solvents evaporate, becoming volatile organic compounds (VOCs) that are known to cause respiratory and neurological problems (water-based paints may be 'low VOC' but they contain additional chemicals that are far from benign). Two weeks later, the paint is still fresh enough to give Dorothea a good initiation into a lifetime of indoor air pollution.

Not surprisingly, synthetic paint manufacturers don't advertise their ingredients, so if you want this information you will have to look elsewhere, to the natural paint suppliers. Natural paints, derived from plants, minerals and the odd insect

## design detail

### 1920s wallpaper

Many modern wallpapers are not made with paper at all but with vinyl and those that are manufactured with paper are rarely sustainably sourced. We therefore decided to avoid using wallpaper except for a feature on one wall of the bedroom. The wallpaper used here is an original 1920s product (not a design reproduction) from a specialist supplier in Sweden. This may not be an economic strategy for decorating an entire house, but it is perfect for a delicate, evocative detail.

(cochineal is made from crushed bugs) have an impressive track record. The remarkable paintings in the Grotte Chauvet in France are more than thirty thousand years old, so you needn't worry about quality and durability.

The paints we are using in Tree House are made by Auro, a German company that only uses natural raw materials, all of which it is happy to declare. Although more expensive than a comparable pot of petrochemicals, there are no hidden environmental costs in manufacture and waste disposal. In fact we can compost any paint we have left at the end of the job.

If you have the time and enthusiasm, you can always make your own paint. It's amazing what you can do with milk and lime juice, the basis of casein paints. Bear in mind that natural paints are not identical to synthetic paints in their use and application, so don't force them on an unsuspecting decorator unless she or he has used them before. Get advice first from a good eco-merchant.

There will be no pinkwash, greenwash or whitewash in Tree House. But you must forgive me for the odd lapse into gingerwash – don't let anyone persuade you that cats are anything other than 100% eco-friendly.

## resources

All the paints and oils for Tree House were supplied by Auro (01452 772020, www.auro.co.uk).

For advice on using and sourcing natural paints and finishes, get advice from a good eco-merchant such as The Green Shop (01452 770629, www.greenshop.co.uk) or Construction Resources (020 7450 2211, www.constructionresources.com).

### Design detail
Vintage 1920s wallpaper from Interior 1900 (www.interior1900.com).

### Publication
*The Natural Paint Book* (L Edwards and J Lawless, Kyle Cathie 2002). An attractive, detailed guide to making and using natural paints.

# 30 NOV 05

By the end of November, most of the second-fix electrics were complete and we were able to turn the lights on for the first time. This was a special moment: another ineluctable step towards completion. Unfortunately not everything was progressing so smoothly. In particular, our heating system languished in bits, awaiting the attention of a plumber who was confident with solar panels, heat pumps and ground loops. As we were stitching these components together in a very unusual way, this took much longer to resolve than anticipated.

Without a heating system, we could not bring the house up to temperature and begin the process of timber stabilisation. We had decided to wait until the end of the first heating season (i.e. the following spring) before laying our salvaged parquet floors in order that both the floors and the house could adjust before we stuck them together. We wanted to move faster on the teak strip for the bathroom as this room had to be completed before we moved in but, with no heat at all in the building, we had to bide our time.

Our cylinder and heat pump are put in position, but the connections remain to be made.

On the roof, architect and site foreman make sure that our man from Lambeth Building Control is happy with progress.

Dave fits one of many compact fluorescent downlighters.

# The low-powered home

I've had enough of being environmentally friendly. In fact, I'm fed up with the whole idea. It's really rather last century don't you think?

Don't worry, I'm not writing this from a 4x4 on the road to Damascus. I'm merely questioning the usefulness of the term 'environmentally friendly'. For starters, it is hopelessly vague, meaning anything and nothing and often disguising a multitude of sins. More importantly, it is just too wet. A little extra 'environmental friendliness' will not save the world from ecological catastrophe, especially if this appears to be no more than one consumer choice among many.

If we are to make a radical difference to our environmental impacts, we need to build and design with radical intent. The influential book *Factor Four* argued that four-fold improvements in the environmental performance of products and buildings are already well within our grasp. All we need is the ambition.

The area where this matters most is energy. If the speed of glacier flow in Greenland has increased by a factor of 100, we need to up our game fast. Although every environmental impact of Tree House is of concern to us, our greatest efforts have been focussed on reducing energy consumption, hopefully to a level that we can provide for using our own on-site renewable supply.

Most of the energy burnt in UK homes is for space heating, so we have put lots of effort into our insulation and airtightness. But as electricians Simon and Dave complete the internal wiring in Tree House, I have been reviewing our plans to keep our electricity consumption as low as possible too.

When people are asked to identify the principal source of electricity use in their homes, the most common reply is the washing machine. In reality, this is some way down the list. The really big burner is lighting. Here factor four is easy: all the lights we are installing in Tree House are compact fluorescents, every one of which achieves at least a factor five improvement over its tungsten equivalent (7W instead of 35W, 11W instead of 60W). Thanks to the work on bulb miniaturisation of our supplier, Megaman, we will be using this technology not only for globe bulbs of all sizes but also for candle bulbs and ceiling recessed downlighters (instead of halogens).

The energy efficiency of kitchen appliances has improved considerably over the last ten years but the worst fridge on the market will still use twice the electricity of the best. And you'll get a factor 160 improvement for using a washing-line instead of a tumble dryer (hanging out the washing uses about 0.025 kWh of your energy vs. 4 kWh for the tumble-dryer).

Regrettably electronic gadgets such as televisions, audio and computers rarely have energy labels so there is more scope for screwing up here. Do you want a plasma screen when it uses nearly twice the energy of the alternatives? Do you want an amplifier that makes a virtue of the size of its transformers? Above all, do you want a desktop computer when it consumes ten times more energy than a laptop?

A few lightbulbs and a laptop computer might not seem like a radical agenda for change but if everyone paid attention to these details we could abandon a clutch of power stations and stop worrying about new nuclear build.

Environmentally friendly? Bah, humbug! We take and take and the environment gives and gives. Our long-suffering friend may be losing patience, but if we are honest to ourselves, there may still be time to save the relationship.

## design detail

### CASSIA door lights

Compact fluorescent lamps always run off mains voltage, so you cannot use them as direct replacements for low-voltage lamps. This was only a problem for us in one instance: the copper up/down-lighters at either side of our front door. These lights were designed for low-voltage halogens which, given the relatively short periods when the lights will be on, would have been perfectly adequate for the job. However we decided to go one step further and replace the halogens with low-voltage LED alternatives, which consume only 1W per lamp instead of the 30W for each of the halogens.

*profile*

## Simon and Dave, the electricians

There are rather a lot of wires in Tree House. As well as power and lighting, we have a CAT5 network, the energy monitoring network running in parallel with the power circuits, the power cables from our photovoltaic roof and a few audio lines thrown in for good measure. Simon and Dave, the electricians, thought nothing of all this and got on with the job with the kind of speed and efficiency to delight contractor and client alike.

*resources*

The compact fluorescent lighting for Tree House was supplied by Megaman (0845 408 4625, www.megamanuk.com).

The Energy Saving Trust has an extensive database of energy-efficient products (0800 915 7722, www.est.org.uk).

### Design detail
CASSIA lights supplied by Light Projects (020 7231 8282, www.lightprojects.co.uk)

### Publications
*The Energy-Saving House* (T Salomon and S Bedel, Centre for Alternative Technology 2003). Examines all the energy issues within the home in detail.

*Factor Four: Doubling wealth, halving resource use* (E von Weizsacker, AB Lovins and L Hunter Lovins, Earthscan 1998). Seminal book on eco-efficiency.

*Natural Capitalism: The next industrial revolution* (P Hawken, AB Lovins, L Hunter Lovins, Earthscan 1999). Takes the factor four argument further, making the case for factor ten improvements in energy and resource use.

# 07 DEC 05

As the year lurched towards the great festive wind-down, the guys on site were beginning to talk about where they would be working in the New Year. This appeared to me to be somewhat premature as it seemed perfectly obvious where they would be working.

Although we were making steady progress, there were some substantial tasks still to be tackled, notably the staircase, heating system, bathroom and pond. At least the list no longer seemed endless. We might be stuck in our Brixton flat for yet another month in the New Year, but completion was no longer the stuff of dreams. The butterfly was beginning to emerge from its cocoon.

*For his last job on the roof, Steve fits leaf guards to protect the hard-to-reach gutters.*

*Under the deep eaves, Steve and Genc work on the balcony subfloor.*

# External shading

One of the joys of living in the country (allegedly) is a heightened awareness of the passing seasons. Which is not to say that we urbanites fail to notice the frost, only that we easily miss the subtleties of the changing flora and fauna around us. Central London may not be devoid of nature's gifts but they get pushed into the background by a surfeit of bricks, mortar and tarmac.

Tree House has been designed to redress the balance. In our open-plan study at the top of the building we have a private view of the seasons through the floor-to-ceiling windows that open on to a balcony right in the crown of our tree. Sure enough, in my regular morning visits to the site I have been struck not only by the longevity of the leaves this year but also by the subtly changing light as they fade from deep summer green to a translucent yellow, variegated by the cracked browns of the wind-burnt leaf edges.

*External blinds keep the heat out as well as controlling glare.*

The study faces east, so as the leaves fall more sunlight penetrates the room, warming the house at the beginning of the day. The tree's timing is impeccable: lifting its shade when the temperature drops, then restoring it in spring to prevent the room from overheating in the summer.

External shading has never been a high priority in the design of British homes but with rising summer temperatures this ought to change. Forty thousand domestic air-conditioning units were sold in 2004, a 27% increase on sales in 1996. Recent

work by the University of Manchester has highlighted the alarming possibility that hotter summers could lead, by 2050, to carbon emissions from cooling outstripping all other domestic emissions put together. Do we want to end up like New York, which ground to a halt when summer electricity demand for air-conditioning tripped the entire power supply to the city?

This scenario can be avoided if we design buildings to stay cool passively. Good natural ventilation across or up a building can remove warm air quickly and exposed heavyweight materials, such as tiled floors, will smooth temperature swings across the day by absorbing and slowly releasing heat. But the most robust strategy is to stop the heat getting inside in the first place. Better insulation, especially in the roof, will help to keep heat out in the summer but you still need to control the impact of direct sunlight pouring through windows. External shading performs this function well but is tricky to get right because daylight and winter solar warmth both make a valuable contribution to a low-energy house.

In the top room of Tree House, the tree's excellent season-specific shading is complemented by deep eaves that will shade the high summer sun but let through the low winter sun. All our windows have deep external surrounds for the same reason. On the ground floor, the glazing that runs the width of the living space will be protected by exterior Venetian blinds and, where the view is less critical, a fixed framework of horizontal wooden slats (a *brise soleil*). There will also be a pergola beyond the windows over which we will train a deciduous climber. External shutters are widely used in southern Europe to keep the heat out during the siesta but as we don't plan to have an afternoon snooze every day we have stuck to design strategies that will help keep the house cool but bright.

Alas, our big bright study is not yet graced with books and furniture but with dusty piles of timber and insulation offcuts. As another completion date slips by with plenty still to do, I feel remarkably calm. The winter will pass; the spring will come. The tree will flourish again and we will be here to watch every bud unfurl.

*The view from the study, just before the final leaf fall.*

## Second-hand curtains

Curtains do little to reduce the risk of summer over-heating because they are on the wrong side of the glass but they do a great job keeping the heat in during the winter. We have curtains over the two glass sliding doors in our living space and study, the only areas of double-glazing in the house as all the fixed windows are triple glazed. Like most aspects of home

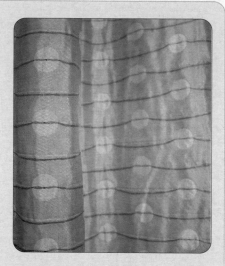

furnishing, high-quality curtains can readily be bought second-hand.

The external Venetian blinds for Tree House were supplied by Lynn Westward (020 8742 8333, www.lynnwestward.com).

### Design detail
Second-hand curtains from The Curtain Exchange (020 7731 8316, www.thecurtainexchange.net).

### Publications
*Daylighting Design in Architecture: Making the most of a natural resource* (Energy Efficiency Best Practice Programme ADH011). Free to download from the Carbon Trust (www.thecarbontrust.co.uk).

*Reducing Overheating – a designer's guide* (Energy Efficiency Best Practice in Housing CE129). Includes sections on movable and permanent shading. Free to download or order from the Energy Saving Trust (0800 915 7722, www.est.org.uk).

# 14 DEC 05

The removal of a substantial part of the scaffolding was a very welcome Christmas present for clients and builders alike. At last we could get a good view of the outcome of so many months of hard labour – and no-one was disappointed. Steve and George had now been on site for fifteen months; Pete, Genc and Nikolin since the frame went up in the hot summer months. Each had played an important part in creating something that we could all now see was going to be quite special.

So we didn't make it 'in for Christmas' as predicted with such confidence back in mid-December. But Ford and I did have our Christmas lunch on a makeshift table in the top room of Tree House, under the great Douglas fir trusses still reeking of oranges. We were happy in the knowledge that there would be many more Christmases to enjoy in this ever more magical space.

*Steve takes a last look at the roof before the scaffold comes down.*

Team Tree House in front of
their emerging craftsmanship.

The tree trunks for the
staircase arrive from Sussex.

# Energy rating

The end is nigh. Or so it seems, as the scaffolding team returns to Tree House and the building starts to emerge from behind its metal curtain. For the first time it is possible to take in the full arboreal form of the house: the vertiginous cedar-clad trunk, the branching Douglas fir of the roof structure and the muscular oak-framed windows that move up the face of the house as our tree's bare limbs rise to frame the cold winter sky.

This project has not been without its problems but in time they will all become shadows in the bright light of a creative ambition realised. There's still plenty to do before we move in but tree and house feel increasingly like equal partners: two striking timber structures, exposed to the elements, drawing their energy from the sun and offering protection to the creatures that seek refuge under their canopies.

Of course we still don't know how the house will perform in practice, especially whether or not we achieve our 'zero carbon' goal. But this week I did have the chance to get our figures checked out by the experts at NHER (National Home Energy Rating). To my relief, their results were remarkably similar to the sums I did long ago before the house was built.

From June 2007 a home energy rating will be required in every home seller's information pack. This rating will be calculated using a simplified version of the Standard Assessment Procedure (SAP) that is used to satisfy the energy-related building regulations. The SAP is currently changing its focus to carbon emissions so from next year a new house ought to emit 20% less carbon dioxide than a house built to today's regulations (an average of 2.6 tonnes per year instead of 3.2 tonnes).

One way to reduce carbon emissions is to install on-site renewable energy. NHER are using Tree House as an example of how an efficient and airtight building fabric can be combined with renewable energy to achieve a very high SAP rating. The new SAP runs from 1 (hopeless) to 100 (no energy bills), with anything over 100 reflecting a net income from energy generation. Although Tree House came out at a disappointing (if excellent) 94, this reflects an assumption that the price we get for electricity we sell to the grid will be less than half the price of what we buy back. In reality most current deals are far better than this. On purely energy terms, we should come out with a 7% surplus.

If you want to get a simple energy rating for your own home, there is a DIY calculator on the Energy Saving Trust website. If you want to pay someone to do the bells and whistles for you, contact NHER. It's a great way to get a really detailed picture of your scope for saving energy.

Ask me in a year's time how all these calculations fare in practice. Right now I am hopeful, inspired by the emergence of a building that is no less exciting than our extremely well-thumbed architectural drawings.

If I am moved by the flowering of the house, this is because in the moment of creation – of a house, a human being, a universe – another point in the future is necessarily defined: the moment of return to nothingness. This is not an unhappy thought ('everlasting life' is for me the ultimate oxymoron) but it is a sobering one. I noticed this week that the Green Party has a new leaflet illustrating 'The British Isle' post-Arctic meltdown: a shrivelled, emaciated remnant of our green and pleasant land. Needless to say, Clapham is not on it.

But the end is not nigh. It may take all our creativity to keep it that way, but 'tis the season to be hopeful and I plan to keep my optimism alive for at least the next fifty years.

## The model

Do you think in two dimensions or three? Can you look at a plan of a room and imagine yourself walking round it?

During the design process for Tree House, our grasp of the project's potential took a big leap forward when our architect, Peter Smithdale, made us a very dinky 1:50 model of the house, complete with timber cladding. Not only did it give us a sense of the bulk of the house, we could also take it apart floor by floor and examine the potential of each room. Although the model has been a bit bashed about, it is still cherished within the walls that it so helpfully prefigured.

Tommy the cat was surprisingly respectful of our dreams and only knocked the roof off.

Try the DIY home energy rating on the Energy Saving Trust website (www.est.org.uk).

For a detailed home energy assessment, contact National Home Energy Rating (01908 672787, www.nher.co.uk).

Your energy supplier has a responsibility to help you improve the energy efficiency of your home and may offer a self-complete survey within a wider package of measures.

EcoHomes is a rating scheme with a broader scope, covering energy, water, pollution, materials, transport, ecology and land use, and health and well-being (www.breeam.org/ecohomes). Only a qualified assessor can give you an EcoHomes rating but you can estimate your home's performance using the pre-assessment estimator, available on the website.

# *11 JAN 06*

The New Year began with the arrival of our front door, a bespoke piece of joinery designed to merge with the cladding details on the front of the house. The verticals didn't quite line up at first but, with a little jiggling of the cladding, all was well.

As the end approached, more and more jobs fell to Ford and me, partly because they were jobs we felt able to do, like decorating and the interminable teak-cleaning, and partly because we had no money left. The physical demands of all these tasks inevitably added to the sense of exhaustion that was now beginning to take hold of me.

*Nik and Genc put our bespoke front door in its long-awaited place (left). Cleaning the teak floors was never one of Will's favourite tasks (above).*

## Composting

We have just moved into our fabulous new home in Clapham. It's a beautifully made timber construction where we can relax with our friends in warmth and comfort and enjoy a regular diet of organic food. Although we're as smug as Christmas, our guardians don't appear to be quite as cheerful. But we don't care about them. After all, we're just a bunch of worms.

Am I really jealous of a box of worms? Well, no. But it is a rather attractive box and they do seem to be particularly happy in it. Ford and I, on the other hand, are still awaiting the day when Tree House is finally fit for human habitation. Although we did enjoy Christmas lunch in the striking top room of Tree House, it was a picnic set up on a carpenter's table. The house is undoubtedly looking great but we still lack a few home comforts including a heating system, a bathroom and a staircase. I'm hoping these will all be in place by the beginning of February but, even at this late stage of the project, I am not counting my chickens.

*Tree House and worm house. Ford welcomes the new residents.*

Which brings me back to our recycling plans. Chickens are a great way of dealing with kitchen waste but the size of our tiny London building plot makes this practically and socially inappropriate. The obvious alternative is a compost heap.

We currently have a plastic compost bin in our back yard, the type that looks like a decapitated dalek. Like most urbanites, we are not experts in the dark arts of composting and simply want a reliable method with minimum hassle. We have learnt that kitchen waste does not contain enough carbon and must be supplemented by toilet rolls, egg boxes and torn up cardboard if a mass of slime is to be avoided. Beyond this, we feed the beast, fork it over occasionally and hope for the best.

*The front door in place with its Douglas fir and cedar details picking up the pattern of the cladding.*

The process works but it is very slow and our dalek is unforgivably ugly. So I have been looking for something that will do the job faster, more discreetly and with a tad more elegance. Our hand-made wormery from Bubble House Worms in Worcestershire meets this specification perfectly. The great thing about a wormery is you don't have to wait around for your carrot peelings to perform their miraculous transformation back into crumbly earth. The worms do the job in a fraction of the time, chomping away without complaint and producing a consistent, nutritious feed for your pot plants. A quicker throughput means that much less space is required, so the big black bin can be replaced by a smaller, smarter box.

Our wormery has two cabinets side by side. Once the worms have eaten their way through the contents of one, they migrate to the other, leaving you to refill the first. Bubble House also make stacking versions, which the worms crawl up, leaving their wormcasts behind for you to harvest. They even have a model with a herb garden growing on top.

Composting is a domestic eco-essential. Not only does it keep valuable organic resources in use, it also stops kitchen waste turning into methane in landfill sites (methane has 21 times the global-warming potential of carbon dioxide). If you have been put off composting because of a lack of space or worries about the stinkiness of rotting vegetables near to your back door, try a wormery. In my humble opinion, worms have the edge over daleks in both efficiency and style.

When I began this column in September 2004, I never dreamt that I would still be writing it in January 2006. I have got so used to being a visitor to Tree House from two blocks away that the idea of actually moving in seems faintly absurd. Hopefully, when the fabled moment finally arrives in a few weeks time, we will be enfolded by the house like disorientated worms in a bed of fresh potato peelings.

*The worms hard at work.*

design detail

## Bath cradle

Our determination to create a home that produced a minimum of non-recyclable waste was not entirely matched by our efforts to reduce waste during

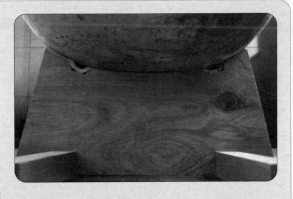

the construction of the building – rather more skips were carted off than I had expected. Nonetheless a good deal of scrap timber and metal was retrieved and reused. The cradle for our salvaged roll-top bath was made out of two chunky pieces of cedar, off-cuts from the staircase, and Douglas fir poles left over from the balcony balustrade.

resources

The wormery for Tree House was supplied by Bubblehouse Worms (01886 832559, www.bubblehouseworms.com).

Basic advice and information about composting is available on all the major recycling websites including www.wasteonline.org.uk, www.letsrecycle.com and www.recyclenow.com. A more detailed guide to the different approaches to composting is provided by the HDRA (www.gardenorganic.org.uk).

The Community Composting Network provides advice and support to people running or planning a community composting scheme. (0114 258 0483, www.communitycompost.org)

### Design detail
Made-to-measure bath cradle by Katie Keat, furniture restorer and cabinet-maker (www.roseberryhouse.co.uk).

### Publication
*Composting: an easy household guide* (N Scott, Green Books 2006). A clear and concise overview of all aspects of composting.

# 18 JAN 06

Scaffolding is a powerful disguise: when the very last boards and poles were stripped from the front of the building, the effect was overwhelming. There before us, in fully fledged three dimensions, was a vision that had occupied my mind's eye for years.

With the outside now looking bright, white and tidy, the inside looked even more of a mess. There was still a lot of joinery, plastering, electrics and plumbing left to do and every task seemed to leave a trail of rubbish and dust in its wake. Trying to finish one space while storing materials for another meant that boxes were constantly being moved around the house from one temporary storage point to another. The amount of *stuff* generated by a building site never ceased to amaze me.

# Building form

January is a puritanical month. If only we scheduled our midwinter festival for February instead of December, we could recover in the knowledge that spring was just around the corner. Instead we find ourselves in the cold dark days of January with nothing but haggis-slaying to look forward to.

If nothing else, these are ideal conditions in which to review the thermal performance of our homes. The domestic experience of warmth and comfort is affected by factors such as insulation, draught-proofing and the efficiency and controls of the heating system. If these are not doing a good job, they can be upgraded, but there is one important building characteristic that you can probably do little about: the shape of your home.

The shape of Tree House was finally revealed in full this week when the last of the scaffolding was removed. It is certainly striking, with its curved timber-clad stair tower, top floor balcony and big southern-pitched solar roof. Despite these details, however, it is basically a big square box. This is a good shape for energy efficiency as the compactness of the form minimises the number of exterior surfaces through which heat can escape.

It is fascinating to look at the shape of buildings and consider how appropriate they are to their climatic conditions. A sphere has the lowest external surface area for a given internal volume so it is no surprise to find hemispherical and conical buildings in very cold climates where preservation of heat is a priority, such as igloos, yurts and wigwams.

At the other extreme, the worst shape for a building in a cold climate is a cross as all the walls are exposed to the elements. Churches have a deserved reputation for being chilly buildings in Britain but are welcomed as cool sanctuaries in Italy. The first building I ever worked in was Charing Cross Hospital in west London, a cruciform tower block that gobbled more energy than a sizeable town.

If you have the money and the inclination, you can ignore these issues and simply burn more fuel. Although British country houses were originally fairly compact in form, the increasing wealth of the ruling classes was expressed in far-flung wings reached by long, exposed corridors. A similar profligacy can be seen today further down the social hierarchy in the addition of large heated conservatories to otherwise compact houses. Such extensions add lots of exposed wall and roof to a home and, as they are made from glass, chuck away heat with abandon.

The homes with the fewest exposed walls are flats and terraced houses, sandwiched between neighbours who hopefully give as much heat as they take. Although Tree House is detached, it may

*The form of the building is simple but striking.*

one day be the first house in a terrace: the northern wall is window-free, designed to enable our neighbours to build right up against us if and when they choose to.

For simple lessons in the relationship of form to thermal performance, get a cat. I am writing this next to three nearly spherical balls of fur. Cats are very sensitive to temperature and adept at changing their shape to suit. Come the summer, these three balls will unravel and stretch out on the cool slate ground floor of Tree House, exposing their bellies and maximising their heat loss.

If you are a puritan at heart, follow the example of medieval monks whose exposed cloisters are as cold in the British climate as the churches next to them. If not, I recommend the compact feline approach to winter living, driven by the devil's instinct for total comfort and idleness.

*The architect's early 3D projection*
*of the front elevation of the house.*

*The first view of the house without scaffolding.*
*The vision has been realised without compromise.*

## Stair tower window

The basic form of Tree House may be a big white box but this is transformed by the detailing, inspired in so many ways by our tree. Half-way up the trunk-like stair tower is a narrow window giving a striking view on to the tree. By carrying the smaller timber verticals of the cladding over the face of this window, the impression from within is of viewing one tree from the heart of another.

*resources*

For advice on all aspects of the thermal performance of your home, contact the Energy Saving Trust (0800 915 7722, www.est.org.uk).

### Publication
*Shelter* (L Kahn and B Easton, Shelter Publications 2000). Hundreds of vernacular buildings, each with a form appropriate to its climate.

*The rear windows follow the pattern of Thomas Cubitt's
Clapham Park mansions, one of which originally occupied our site.*

# 25 JAN 06

The ladders may have gone from the exterior of the house but they remained depressingly inescapable on the inside. The staircase had been contemplated, drawn, planned and scheduled so often that it was beginning to take on a mythical status. This was in part due to the complications of the design: the newel posts were two debarked tree trunks and the beginning of each flight had to wind round and up the curved form of the stair tower.

When Steve and Nik finally got to work on the staircase, the swear box started to overflow. They were both skilled carpenters, determined to make a good job of an unusual design, but their ideas about how to achieve this were not always convergent. Happily this Laurel and Hardy caper did not result in a fine mess but in a particularly well-made piece of bespoke joinery.

*Steve at work on the staircase, winding up and round a bare tree trunk.*

*Nik fits the treads and risers into the exposed string.*

*The second flight was even more challenging, not least because the newel posts (the tree trunks) were much thinner.*

# Craftsmanship and quality

We have been doing rather a lot of painting this week, creating bright white walls with sweet-smelling natural emulsion, but I keep thinking of wallpaper: richly patterned William Morris wallpaper. This is not because I have a sudden urge to cover our walls with the stuff but because I feel, somewhat indulgently, that Morris would have approved of Tree House.

William Morris founded the Arts and Crafts movement, a late nineteenth-century response to industrialisation and mass production that promoted traditional skills and craftsmanship. Where division of labour stripped craftsmen of their creative involvement in production, the Arts and Crafts pioneers sought not only to hold on to this involvement but to elevate the craftsman to the status of an artist.

Tree House is a highly crafted building, built by a skilled team of craftspeople whose commitment to quality has kept the project in the realm of dreams rather than nightmares. This week, the glittering stained glass panel created by artist Sarah McNicol was installed at the centre of our kitchen, framed by beautiful beech shelves made by cabinet-maker Stephen Edwards (see pp 130–134). Carpenters Steve Archbutt and Nikolin Deda began their final feat of bespoke on-site joinery: the curved staircase that winds round and up two Douglas fir tree trunks. Off-site, contractor Martin Hughes and architect Peter Smithdale have redoubled their efforts to ensure that Tree House emerges as an integrated work of art, craft and engineering.

The exposed trusses in our top room have a medieval character that Morris would have loved and everywhere we have turned to simple, natural materials to bring character to the interior. Our use of stained-glass in a domestic context was pioneered in the Red House, the ground-breaking home built for Morris in Bexleyheath by Philip Webb. We have even treated the garden as an extra room of the house in the tradition of Gertrude Jekyll.

But if Tree House boasts some of the qualities of the Arts and Crafts movement, does it also suffer from its weaknesses? Morris and his followers are often criticised as well-heeled idealists whose pursuit of craftsmanship never reached beyond the pockets of the rich. Is a highly crafted bespoke project such as Tree House just as elitist with little relevance to the challenges of mass house-building?

Only if you take a very narrow view of craftsmanship. As well as a beautiful kitchen and staircase, Tree House has super-insulated walls, high performance windows, an exceptionally airtight outer envelope and three integrated renewable energy technologies. If any of these components had been built or installed

carelessly, as they often are on British building sites, we would stand little chance of achieving our tough 'zero carbon' energy goal. If craftsmanship is seen as a commitment to quality in all aspects of construction, it should be a priority for any eco-build, regardless of budget, from day one.

Craftsmanship may not be a sufficient condition for eco-building (the Red House turns its windows to the chilly north) but it is a necessary condition. Unfortunately quality remains a huge challenge for an industry with a history of building houses with cheap heating systems that disguise all manner of failures in the building fabric. A move towards greater off-site prefabrication may help to address this problem in mass housing but nothing beats skilled and committed people on site.

Morris recommended having nothing in your home that you do not know to be useful or believe to be beautiful. He may have been a crusty old snob – and I'm not convinced that his own wallpaper consistently meets these criteria – but this is a worthy maxim for our careless, throw-away world.

| *design detail* | Contemporary Chinese ceramics |

As my father worked in the Far East, I grew up among walls lined with traditional Chinese craftsmanship including many ornamental plates. His style is too traditional for Tree House but it is one of the seeds of my enthusiasm for the contemporary ceramics of Mak Yee Fun. Her work captures the formal purity of traditional Chinese ceramics but replaces

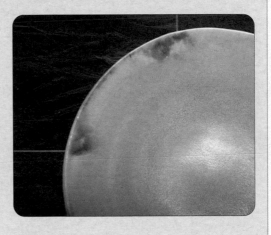

graphic complexity with a subtle density of light and colour.

*The Douglas fir tree trunks rise up through the house, supporting the staircase.*

## Calvin Plomer-Roberts

Calvin the plasterer was one of the most dedicated craftsmen who worked on Tree House. His finest hour was the creation of the beautiful curved ceilings that wind up and round the tree-trunk newel posts of the staircase. Pure Gaudi.

The William Morris Gallery is sited in Morris's home in Walthamstow, north London. It includes displays of furniture, textiles, ceramics and glass by Morris and his followers in the Arts and Crafts Movement (020 8527 3782, www.lbwf/wmg).

The Red House is now owned by the National Trust and is open to visitors (01494 559799, www.nationaltrust.org.uk).

### Design detail
Ceramic dish by Mak Yee Fun (yeefun@ma9978.freeserve.co.uk).

### Publications
*The Beauty of Craft* (Edited by S Brown and M Kumar Mitchell, Green Books 2004). A fascinating international anthology of craftsmanship with a scope ranging from knitting to architecture.

*Places of the Soul: Architecture and environmental design as a healing art* (C Day, Architectural Press 2003). A plea for sustainable design that encompasses the craft of building and the creation of exceptional spaces.

# 1 FEB 06

By the beginning of February the finer details of the project were beginning to shine through. These included the kitchen with its glittering central panel of stained glass, the balcony balustrade clad in timbers that rise and spread in a Fibonacci sequence and the front fences, evoking the delicate density of the tree's form. These were much more than finishing touches: each had a long-considered and long-awaited part to play in the whole design.

Not that we were anywhere near finishing. The 'to do' list seemed endless as every time something was crossed off, something else popped up to take its place. Our determination to move in created a slight sense of urgency on site but it was increasingly clear to us that our first experience of living in Tree House was unlikely to be comfortable.

# Eco-frugality

I bought a ticket for the £100 million lottery last week. Not out of desperation mind you. The construction of Tree House has cost more than anyone anticipated but we can still afford to live there without selling our souls. What's the point of a good credit rating if you don't make full use of it, I say.

*Stephen Edwards and the almost complete kitchen.*

I bought the ticket simply to participate in the national contemplation of limitless opportunity. Perversely, in this existential moment I found myself recoiling from the thought of spending any money at all. To a degree this is because we have been shedding so much cash over the last sixteen months that I am rather keen to stop for a while and enjoy the rewards of all our spending. But more profoundly, I realise that my ambitions post-build are remarkably frugal.

Tree House has been a big investment of money, resources and energy, but when we're living there (in three weeks time) we will be self-sufficient in energy and very thrifty in our use of water and other resources. Without doubt it will also be a

delightful place to live. Our newly completed kitchen, a work of bespoke joinery by local designer-maker Stephen Edwards, expresses our ambitions for the house very concisely. It includes low-impact materials, appliances that consume a minimum of energy and water, low-energy lights and lots of built-in recycling capacity. But above all it's dead gorgeous, designed to make cooking a joy. If there is such a thing as 'contemporary frugality', perhaps this is it: a wonderful, low-impact space that we won't want to leave, however many high-energy restaurants beckon from Clapham High Street.

*Jonnie Rowlandson installs the front fence.*

My lottery ticket has made me aware of how fully I want to explore this conjunction of delight and eco-frugality in all aspects of my life. For example, since I stopped shopping in supermarkets I have become a regular visitor to Pimlico Road farmers' market and have discovered the specialities of local shops that I previously ignored. Shopping may take longer, and sometimes cost a bit more, but my food miles are radically reduced, the local economy benefits, the produce tastes better and the experience is a genuine pleasure.

Another key decision has been to get rid of our car once we have moved in. Having recently started cycling again in preparation for this moment, I have been struck by the pleasures of cycling in London's back roads, especially on bright winter days when the richness of the capital's architecture surprises you on every street.

More radically, I have decided never to fly again. This choice may narrow the geographical range of my travel options but are there really any earthly pleasures that cannot be enjoyed within the bounds of Britain and Europe? And shouldn't travelling itself be a pleasure rather than an infernal ride in a dehydrated sardine tin? (If our friends in Sydney are reading this, don't worry – we'll get there eventually, albeit via a much more interesting route than most people ever contemplate.)

*The balcony balustrade grows out of the cladding of the stair tower.*

I have always insisted that there are no hair shirts in my wardrobe and that Tree House will give us everything we want from a home. Given these wider decisions, you may suspect that I have been wearing hessian underpants all along. Perhaps you are right. Or perhaps I have been awakened to a basic reason why people have always chosen to live more simply: life's just more interesting that way.

So what will I do with the £100m, given that I won't be rushing out to buy fast cars and a private jet? Easy: I plan to buy everyone in London a packet of courgette seeds, a guide to the National Cycle Network and a day return to the seaside. Risky, though: if all seven million of us have a glut of marrows next August, there'll be hell to pay.

## design detail

### Trevor

Cats understand very well that a frugal life need not be a harsh life. After all, what else is there but food, love and a place in the sun?

## resources

The kitchen for Tree House was designed and built by Stephen Edwards (020 7737 8110 www.ecointeriors-uk.com).

### Publications
*A Handmade Life* (W Copperthwaite, Chelsea Green 2003). A practical and spiritual vision of simple living.

*The New Complete Book of Self-Sufficiency* (J Seymour, Dorling Kindersley 2003) An accessible guide to life beyond consumer culture.

# 15 FEB 06

Even with relatively few rooms, it's amazing how many internal surfaces there are to finish in a medium-sized house. We knew by now that many of these surfaces would still be unfinished when we moved in, so we focussed on the areas that mattered the most: laying teak boards in the bathroom so that the sanitary ware could be installed, painting the room at the top of the house before a mass of boxes made this task impossible and tiling the living room and kitchen floor, in the hope that we would not have to wait too long before these spaces became accessible.

On top of all this, there were neglected external surfaces to tackle. In the front garden, we wanted a porous surface that would soak up the rain. In the back we needed precisely the opposite: the pond had to be tanked very thoroughly before any of the increasingly rare rainfall would stay in it.

The pace of work on site was now slowing, with many familiar faces leaving for good. It was sad to see people go, especially those who had contributed much to the project, but it was also reassuring to see that the jobs on site were genuinely beginning to run out.

*Calvin climbs out of the pond to consider how the last corner can possibly be tanked.*

*Ford oils the newly laid reclaimed teak floor in the bathroom.*

# Porous paving

If Ken Livingstone announced that he wanted to pave over Hyde Park to make space for a gigantic bus depot, the response would be predictably ballistic. Such a plan seems unlikely, given our dear leader's interest in protecting London's environment, yet in reality the displacement of the capital's green spaces for car parks is relentless – and on a far greater scale. The London Assembly recently found that 12 square miles, or 22 Hyde Park's worth, of front gardens have been paved over for cars in the capital.

This quiet destruction has an enormous impact as our streets become ever harsher places for all forms of life. The loss of biodiversity may not be obvious when one household pours the concrete but by the time several others have followed suit, the effect on animal, vegetable and human quality of life is unmistakeable.

Gardens also moderate the effects of the climate. A garden-free urban environment can overheat in summer because buildings and roads absorb and then radiate heat, often long into the evening. Plants help to cool things down, partly by providing shade but also through transpiration as the evaporation of water absorbs lots of heat from the air. Furthermore, gardens reduce the risk of flooding during storms by absorbing and slowing the deluge, keeping drains underwhelmed. This is especially important in London where most storm drains are connected to the sewers, with disastrous consequences when capacity is exceeded.

Down at Tree House, we want the wildlife garden at the front of the plot to fulfil all of these ecological functions. Unfortunately, however, our planning consent requires us to put in a 'hard standing' in our front garden even though we have no intention of keeping a car there. We are therefore in the unusual position of wanting to let our car park become a garden, rather than the other way round.

Fortunately, it is not difficult to design a space to perform both of these functions. All you need is a surface that will support the weight of a vehicle while also admitting rainwater and allowing plants to thrive. 'Porous paving' ranges from sweeping gravel driveways to webs of brutalist concrete hexagons. We are using a plastic mesh, which

all but disappears when packed with gravel or earth. The Grassington Paving System, comes in interlocking sheets that can be combined to fit spaces of all sizes. We will let creeping jenny,

*Our Eluna leaf tile set within the Grassington porous paving.*

bugle and thyme grow through the gaps, hardy plants that will cope with abuse from our visitors' tyres.

The gaps within the mesh also give us space for a little arboreal allusion, courtesy of the Green Bottle Unit in Hackney. This innovative company makes Eluna paving tiles from recycled glass, successfully exploiting the variegated patina and colour of the material. In the autumn I photographed a fallen leaf from our tree, turned the image into a silhouette and sent it to the Unit to turn into something special. The result: eleven blue-green leaves that will never fade, guiding visitors on the short path through our garden to the door of Tree House.

There is enormous scope for improving our urban environments by expanding the functions of our many hard surfaces. If I were the mayor of London, I would insist that all bus depots become living buildings with earthen roofs, rainwater harvesting and solar power. As I can reasonably claim to be the man on the Clapham omnibus, Ken can be assured that this idea will be met with universal approval.

## *design detail* Mailbox

Our postman does not have to traverse the entire length of our porous paving every morning as we have a wall-mounted mailbox near the gate. By having a mailbox outside the envelope of the building, we avoid the significant losses of energy from a draughty letterbox in the front door.

## *resources*

The Grassington Paving System for the front garden of Tree House was supplied by LBS Garden Warehouse (01282 873333, www.lbsgardendirect.co.uk).

The Eluna recycled glass paviours were supplied by the Green Bottle Unit (www.eluna.org.uk, 020 7241 7474).

### Design detail
Lockable stainless steel mailbox from Safepost.

### Publication
*Gardening Matters: Front Gardens* (Royal Horticultural Society, 2006). A concise guide to protecting and cultivating your front garden while also parking your car there. Available online and as a booklet (020 7834 4333, www.rhs.org.uk).

# 22 FEB 06

The one truly interminable task on this project was the plumbing. The week before we moved in we had no running water, no heating and no working toilet. Only one of these omissions (the last one) had been rectified seven days later.

The domestic plumbing was actually quite simple: we had deliberately located all taps, showers and appliances very close to the hot water cylinder to minimise pipe runs and heat losses. The plumbing for our heating system, on the other hand, took a little more effort. In particular, the unusual way we were linking a heat pump with a solar thermal panel made everyone a little nervous as this aspect of our ambitious design had to be installed exactly right – and no-one had done it before.

The prospect of moving ourselves, our possessions and our four cats into a building site was not a happy one. This time, however, we were not going to back down. It was time for a final push to make Tree House into something resembling a home.

*Steve bids a less than fond farewell to the site portaloo.*

*Mel connects the toilet and bidet in the bathroom.*

# High-tech vs. low-tech

When daily routines are disrupted by stressful events, personal environmental behaviours are often the first to fail. This is famously true at Christmas when the over-flowing recycle bin quickly gets ignored and a mountain of stuff gets bought and immediately thrown away.

Moving house is one of life's most disruptive and stressful activities with a very high attendant risk of overflowing bins. So, as we prepare to move into Tree House, we are doing what we can to avoid a sudden spike in our household's output to landfill, not least by exploiting the excellent environmental resources of eBay and local charity shops. It doesn't help that we are moving into a house that is still being built: if the schedule for this project has sometimes been a little loose, it is now as tight as post-Christmas trousers.

One of the rooms that is unlikely to be 100% complete is our Engine Room, the small room where all our specialist technical kit is housed. Its contents include the inverters that turn the output of our photovoltaic roof into 240V AC, the pump for our solar thermal panel, the ground source heat pump that extracts the sun's warmth from beneath the house and the mechanical ventilation unit that takes the heat from the outgoing stale air and transfers it to the incoming fresh air. There is also a highly insulated hot water cylinder, a communications hub and a big bundle of meters and wires designed to tell us if all this kit achieves our 'zero carbon' goal. Master plumber Mick Nolan is doing his best to connect up the massed pipes and drums but, unfortunately for him, our prior assessment of the space required was not very accurate (Edinburgh Castle it isn't).

Every piece of technology is there for a good reason. Nonetheless I fear this room would turn some eco-builders green with horror rather than envy as it epitomises a high-tech approach in stark contrast to traditional, passive buildings. Some argue that the more gizmos you have, the less ecological your final result is likely to be.

*Mick Nolan and the complex plumbing in the Engine Room.*

I have some sympathy with this position. In one of the more remote valleys of the Scottish Borders there is a house made from cob (packed earth) looking south over a deciduous wood. The walls absorb the heat of the sun, keeping the house warm in winter and cool in summer. Heat is supplied by a wood burner, fuelled by local tree thinnings. The water tank is fed by a stream higher up the hill and a compost toilet is one of the many nutrient sources for the verdant kitchen garden.

This, perhaps, is the house we will build next, if we ever tire of city life. But it is not a specification that suits the centre of London. In Clapham we can only dispense with fossil fuels by combining a very high-performance building with technology appropriate to a small urban footprint.

Eco-technology is problematic when it gets stuck on a building like a flagpole, waving green credentials regardless of what lies below. When technology is used more carefully, as part of a holistic approach to environmental design, it definitely has its place. In particular, I believe that the more we take control of energy generation by doing it ourselves rather than leaving it to distant power stations, the more we will take seriously the challenge of a very low-carbon future.

It will be some time before we establish a new domestic routine in Tree House but hopefully our stress levels will get through this week's peak and we will soon be enjoying our ultra-low impact life in comfort, listening to the birds sing and the gizmos click and whirr.

---

*profile*   Katie Keat

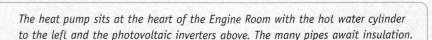

One of the very last jobs to be done was the repair of all our second-hand teak and rosewood furniture which had been bashed, scratched and stained with cat-sick over the course of the build and move.

Local furniture restorer and cabinet-maker Katie Keat came to our rescue, returning tables, chairs and sideboards to their mid-20th-century glory (www.roseberryhouse.co.uk).

---

*The heat pump sits at the heart of the Engine Room with the hot water cylinder to the left and the photovoltaic inverters above. The many pipes await insulation.*

*design detail*

## Dimmable compact fluorescent bulb

This is a gizmo I was delighted to incorporate in Tree House: a compact fluorescent bulb that can be dimmed on an ordinary dimming circuit. As the inability to dim compact fluorescent bulbs has long been an obstacle to their use as direct replacements for tungsten bulbs in homes, this humble bulb is a major advance.

*resources*

A debate on 'eco-minimalism' was published by *Building for a Future* magazine and is available on their archive (www.buildingforafuture.co.uk).

The National Energy Foundation website includes a section for householders that promotes a holistic approach to energy and technology, describing the options for energy conservation, energy efficiency and renewable systems (www.nef.org.uk).

### Design detail
'DorS' dimming series bulb from Megaman UK (0845 408 4625, www.megamanuk.com).

### Publication
*Ecohouse 2* (S Roaf, M Fuentes and S Thomas, Architectural Press 2003). This established guide combines an enthusiasm for technology with a full account of how houses can exploit their climatic conditions passively.

# *1 MAR 06*

The story of Tree House ended exactly one and a half years after it began. The house was far from complete but the transformation of building site into home seemed an appropriate juncture to close the story as well as the front door.

The build was a year over schedule, the final cost – £330,000 on top of £150,000 for the land – was well beyond what we had planned, and I was physically and emotionally exhausted, but Tree House had taken root and was already beginning to blossom.

We set out to build a house with a design and environmental specification well beyond the normal and we learnt the hard way that anything bespoke spells trouble in the building industry. But we have no regrets: in the face of growing environmental crisis, ambition is needed at every level of action, from the individual to the international, and we did the best we could for our little corner of Clapham.

We hope that Tree House, a house that works like a tree, will continue to thrive in years to come, a symbol of the power of ecological design to transform the quality of all our lives.

## Moving in

After three months of drought it poured with rain on Sunday, the day Ford and I finally moved into Tree House. We started work at 7.00am and by mid-morning the short distance from the back of the transit van to our new front door had become a treacherous mudslide. Every precarious traverse across it was followed by a heavyweight push up two flights of stairs to the only habitable room in the building. By nightfall we were flagging badly with no end in sight – the removal fairy kept conjuring up piles of neglected possessions in our Brixton flat whenever our backs were turned.

Just at this moment, when dirt, exhaustion and disorientation threatened to overwhelm us, a passer-by stopped and asked us if we were the owners of the

*The very first view of Tree House as home.*

house. She went on: "It's gorgeous. It's a real statement. I love it being on my road where I can see it every day." We nearly wept into our cardboard boxes.

My first column in *The Independent* appeared on September 1st 2004, exactly eighteen months ago. I knew then that the building of Tree House was unlikely to be straightforward, given how far we had pushed our ecological specification, but if anyone had told me then to prepare for a journey lasting a year and a half I would have scoffed. After all, we were building a two-bedroom house on a small site in Clapham, not some elaborate mansion in the country.

This time last year, six months in, site stalwarts Steve and George were still absorbed with ground works. It was a bleak time and I struggled then to maintain my enthusiasm but such private wobbles never lasted long. Whatever stage the build was at and whatever problems we faced, the tree was always there, strong and graceful, to sustain our vision.

The exceptional ecological performance of the tree provided us with a truly challenging goal for the house, a standard far beyond any current definitions of good practice in the building industry. We cannot claim to have met this standard in every detail but we have done our best and fought off a legion of compromises.

Beyond this ecological goal, at the very heart of our ambition, lies the sheer beauty of our tree. Even in the heart of winter its vaulted form, organic density and depth of texture are an inspiration. It has been exciting and interesting to

*Tree House is revealed in its full arboreal glory.*

build a house powered entirely by the sun but the greatest personal reward lies in the creation of a house that is beautiful.

Very occasionally, when the accumulating evidence of global climate breakdown saps my optimism, I wonder what difference our radical eco-specification will actually make. But I have no such doubts about acts and works of beauty. After all, if we cannot sustain a delight in life itself, whatever future we face, what is it that we are fighting to preserve?

Finally, we must thank the many people – professionals, labourers, craftspeople and artists – whose imagination, commitment and hard work have brought Tree House into being, above all architect Peter Smithdale, contractor Martin Hughes and site foreman Steve Archbutt. Everyone who worked on this job cared about it and, thank God, it shows.

*Tommy has a Rembrandt moment amid the chaos of moving.*

# Index